81 *Module categories of analytic groups*

ANDY R. MAGID

Professor of Mathematics, University of Oklahoma

Module categories of analytic groups

CAMBRIDGE UNIVERSITY PRESS

CAMBRIDGE

LONDON   NEW YORK   NEW ROCHELLE

MELBOURNE   SYDNEY

CAMBRIDGE UNIVERSITY PRESS
Cambridge, New York, Melbourne, Madrid, Cape Town, Singapore, São Paulo, Delhi

Cambridge University Press
The Edinburgh Building, Cambridge CB2 8RU, UK

Published in the United States of America by Cambridge University Press, New York

www.cambridge.org
Information on this title: www.cambridge.org/9780521242004

First published 1982
This digitally printed version 2008

A catalogue record for this publication is available from the British Library

Library of Congress Catalogue Card Number: 81-10215

ISBN 978-0-521-24200-4 hardback
ISBN 978-0-521-09027-8 paperback

Contents

Contents

Preface

I have been interested in analytic groups and algebraic structures on them for some time – ever since an anonymous reviewer of a grant proposal suggested I take a look at some papers of Hochschild and Mostow in connection with the work I proposed. I found the papers fascinating, and useful as I began work in the area, but the foundations of the subject seemed to me (as an autodidact Lie theorist) irrevocably recondite. Then, in the summer of 1978, two things happened: while preparing for a talk for a conference in Copenhagen, I found elementary arguments for the basic existential facts of the theory (these appear here in Chapter 3) and, more important, I met Alex Lubotzby at the conference, who pointed out the connection between the Hochschild–Mostow theory and the Grothendieck theory of Tannakian categories (this appears here in Chapter 2). These things meant that the subject was both easier and of wider application than I had previously imagined, and so inspired this volume.

The book was written while I enjoyed a sabbatical leave from The University of Oklahoma and was Visiting Professor of Mathematics, University of Virginia (Fall 1979), Visiting Professor of Mathematics, Bar-Ilan University (Winter 1980), and Visiting Professor of Mathematics, University of California, Berkeley (Spring 1980). To those institutions and my colleagues there, I am grateful. I especially want to acknowledge the encouragement I have gotten over the years from Gerhard Hochschild.

I am grateful to Trish Abolins for her able and efficient typing of the manuscript, to Hyman Bass for suggesting that the book appear as a Cambridge Tract, and to my family.

Chickasha, Oklahoma   A.R.M.
Thanksgiving, 1980

Notation and conventions

$\mathbb{C}$ denotes the complex field.

Z denotes the integers.

All tensor products are over $\mathbb{C}$.

All vector spaces are over $\mathbb{C}$, and most are finite-dimensional. If V is such a space, then:

$\quad$ V^* is the $\mathbb{C}$-linear dual of V;

$\quad$ $\mathrm{End}(V)$ is the space of $\mathbb{C}$-linear endomorphisms of V;

$\quad$ $\mathrm{GL}(V)$ is the group of $\mathbb{C}$-linear automorphisms of V;

$\quad$ $\mathrm{gl}(V)$ is $\mathrm{End}(V)$ as a $\mathbb{C}$-Lie algebra with bracket $[A, B] = AB - BA$;

$\quad$ $V^{\otimes n} = V \otimes V \otimes \cdots \otimes V$ (n times);

$\quad$ $\langle x_i \mid i \in S \rangle$ is the subspace of V spanned by the subset $\{x_i \mid i \in S\}$.

$\mathrm{GL}_n\mathbb{C}$ is the group of invertible $n \times n$ complex matrices.

$\mathrm{GL}_1\mathbb{C}$ is also denoted $\mathbb{C}^*$.

The diagonal $n \times n$ matrix with diagonal entries $a_1, \ldots, a_n$ is denoted $\mathrm{diag}(a_1, \ldots, a_n)$.

An analytic group is a connected complex Lie group.

The Lie algebra of the analytic group is denoted $\mathrm{Lie}(G)$, which is sometimes abbreviated to $L(G)$.

$X(G)$ is the group of analytic group homomorphisms from G to $\mathrm{GL}_1\mathbb{C}$ (the characters of G).

$X^+(G)$ is the group of analytic group homomorphisms from G to $\mathbb{C}$ (the additive characters of G).

An algebraic group is an affine algebraic group over $\mathbb{C}$, and usually connected.

A torus is an algebraic group isomorphic to a product of copies of $\mathrm{GL}_1\mathbb{C}$.

The affine coordinate ring of the algebraic group G is denoted $\mathbb{C}[G]$.

If V is an affine variety over $\mathbb{C}$, $\mathbb{C}[V]$ is its algebra of polynomial functions.

If G acts on V algebraically, $\mathbb{C}[V]^G$ denotes the G-invariant polynomial functions on V.

If L is a Lie algebra, $Z(L)$ denotes the center of L.

If $x, y \in L$, $\mathrm{ad}(x)(y) = [x, y]$.

If $L = \mathrm{Lie}(G)$, the adjoint representation of G on L is denoted Ad, and if H is a subgroup of G, $L^H = \{x \in L \mid \mathrm{Ad}(h)(x) = x$ for all $h \in H\}$.

If X is a group and H, K are subgroups of G, then (H, K) is the subgroup generated by commutators $hkh^{-1}k^{-1}$ with $h \in H$ and $k \in K$.

$X^{ab} = X/(X, X)$.

$\mathbb{C}[X]$ denotes the complex group ring of X.

e is the identity of X.

If G, N are groups and $s: G \to \mathrm{Aut}(N)$ is a homomorphism, the resulting semidirect product group is denoted $N \times_s G$.

Introduction

1

This book is about the representation theory of analytic groups (an analytic group is a connected complex Lie group and a representation of it as matrices of size $n \times n$ is a holomorphic homomorphism from the analytic group to the group $GL_n \mathbb{C}$ of all invertible $n \times n$ complex matrices). As it is usually viewed, the main problem of representation theory is: given the group, determine, in terms of some 'parameters', the representations of the given group. For example, the classification of representations of a simply-connected simple Lie group in terms of the high weights of irreducible components, and the description of the possible high weights from the root system of the Lie algebra of the group, is a profound and inspiring solution to the problem for the groups to which it applies. (One can get an idea of how inspiring this solution was by consulting, for example, the bibliography of [1]†.)

There is also, however, the converse problem, which will be our major concern: given its representations, determine the group. The problem, as stated, is not very well-posed (what does it mean to be 'given the representations'?) although, as is often the case in the development of mathematics, that was not a serious deterrent historically (the historical development is summarized below), and currently the concepts of category theory allow a precise statement, as we shall see. It turns out, however, that the problem does not always have a solution. There are analytic groups that are not isomorphic, and yet their categories of representations are the same; examples are given in Chapter 1, Example E. Thus our problem becomes: given the representations of an analytic group, what do we know?

† Bracketed references refer to the bibliography.

1

As we shall see, we know quite a bit. To explain the answer, we need to be precise now about the problem. By the familiar mechanisms, we can regard a representation of size $n \times n$ of the analytic group G as giving a structure of G-module on an n-dimensional complex vector space, and vice-versa; see (1.1), (1.2) below.† But by dealing with the G-modules, we can use the additional idea of a G-module homomorphism (1.4) to conceive of the category $\mathrm{Mod}(G)$ of (finite-dimensional, analytic) G-modules. Thus, we interpret the phrase 'given the representations' of G to mean that the category $\mathrm{Mod}(G)$ is known to us, in the sense that it is a specified subcategory of the category of finite-dimensional complex vector spaces. We will also need to know that it is closed under the vector space operations of tensor product, direct sum, and linear dual. This category, while it does not determine G, does determine an (abstract) group, which we denote $\mathrm{Aut}_\otimes(\mathrm{Mod}(G))$, and an algebra $R(\mathrm{Mod}(G))$ of complex functions on it such that the pair $(\mathrm{Aut}_\otimes(\mathrm{Mod}(G)), R(\mathrm{Mod}(G)))$ completely characterizes the category $\mathrm{Mod}(G)$. Thus, we can conclude: given the representations of G, we know $\mathrm{Aut}_\otimes(\mathrm{Mod}(G))$ and $R(\mathrm{Mod}(G))$, and conversely.

Ths neatly answers the question of what one knows when one knows the representations of an analytic group, but does not yet speak of the original problem of recovering G from its representations. It turns out that there is a group homomorphism $G \to \mathrm{Aut}_\otimes(\mathrm{Mod}(G))$, which we also use to make $R(\mathrm{Mod}(G))$ an algebra of functions on G, and using these we elucidate the structure of $\mathrm{Aut}_\otimes(\mathrm{Mod}(G))$ and $R(\mathrm{Mod}(G))$.

The relative logical positions of these two operational modalities (obtaining $\mathrm{Aut}_\otimes(\mathrm{Mod}(G))$ from $\mathrm{Mod}(G)$ as a category, and studying $\mathrm{Aut}_\otimes(\mathrm{Mod}(G))$ via the homomorphism from G to it) need to be kept clear: in the first we are not permitted to glimpse G, but only its category $\mathrm{Mod}(G)$ of representations, while in the second we employ G directly, assuming it known in order to describe its associated group $\mathrm{Aut}_\otimes(\mathrm{Mod}(G))$. The opportunities for confusion

---

†Results, remarks and definitions are so numbered that the digit to the left of the decimal indicates the chapter in which they occur.

are obvious, especially since sometimes it will be necessary to operate in both modes simultaneously, but our general trend is to be as follows: as much as possible, the first will be done first – mostly in Chapter 2 – while the second will be done subsequently. A detailed summary of the book's contents is given in the third section of this introduction. The second section contains a brief history and the fourth contains some general comments about assumed prerequisites.

2

We can begin the story of recovering groups from their representations with Pontryagin's duality theorem of 1934 [26]: here G is a compact abelian group, $\hat{G}$ is the dual discrete group of continuous (complex) characters of G, and $\hat{\hat{G}}$ is the dual group of complex characters of $\hat{G}$. Then his 'first fundamental theorem' asserts that G and $\hat{\hat{G}}$ are isomorphic. A bit of translation is needed to see this theorem in the context we introduced above. First, finite-dimensional representations of G are direct sums of one-dimensional representations $G \to GL_1\mathbb{C}$ which are the characters of G; G-homomorphisms between these one-dimensional G-modules are given by scalar multiplications only, so can be ignored. But the tensor product of these G-modules is what determines the composition in $\hat{G}$. Thus we can regard knowledge of $\hat{G}$ and knowledge of $\text{Mod}(G)$ equivalent in the compact abelian case. The recovery procedure here also deserves some comment: the idea is to define complex functions on $\hat{G}$ (the set of one-dimensional representations of G) which respect the composition rule in $\hat{G}$; that these turn out to be characters of $\hat{G}$ should be regarded as a consequence of the fact that only one-dimensional representations are involved.

This description of the recovery procedure is given with the benefit of hindsight. Once stated this way, the idea of how to extend Pontryagin's procedure to nonabelian compact groups is clearer: again, one can stick to the set $\hat{G}$ of representations of the compact group G, which is now to have the composition rules of direct sum and tensor product of representations, and now $\hat{\hat{G}}$ is to be made of 'representations' of $\hat{G}$, that is, matrix valued functions respecting the composition rules. In this context, Tannaka proved his duality theorem of 1938 [29] which states that G and $\hat{\hat{G}}$ as above are isomorphic.

Chevalley's 1946 book [3] includes a proof of Tannaka's duality theorem, which further identifies the compact real Lie group G with the set of real points of a complex linear algebraic group $\bar{G}$. $\bar{G}$ is obtained by considering the algebra R of complex valued functions generated by the matrix coordinate functions of the representations of G: in contemporary terms, R is in a natural way a finitely-generated complex Hopf algebra, hence, the coordinate ring of an algebraic group $\bar{G}$. The main result is now that $\hat{\hat{G}}$ (the 'representations' of the set of representations of G, as above) is identified with $\bar{G}$.

At this point there are now really two kinds of dual objects to the compact Lie group G: the set $\hat{G}$ of representations of G, with its composition laws from tensor product and direct sum, and the algebra R of coordinate functions from representations, with its Hopf algebra structure. Both objects can be granted equal status if the group is assumed known (they both require knowing the category Mod(G)), but from the point of view we shall be concerned with here, the status of R as a dual object is a bit shaky. It appears that to describe R (it is, after all, an algebra of functions on G) one must know G, not just the category Mod(G). Appearances can be deceptive (we will see in Chapter 2 that R can be produced directly from Mod(G)), of course. Nonetheless, Chevalley's formulation of Tannaka's theorem via R (which came to be known as the algebra of representative functions on G) seems to have led to an emphasis on the algebraically more familiar object R and away from the rather awkward notion of 'representation' of $\hat{G}$.

This is, at least, what Hochschild & Mostow seem to tell us in their 1957 paper [7], the first of a series of ten papers [7, 8, 9, 10, 11, 12, 13, 14, 15, 16] published between 1957 and 1969 on the structure of the algebra $R(G)$ of representative functions on an analytic group G and its applications to the representation theory of G. There were other studies of Tannaka-type duality between 1941 and 1957, by Harish-Chandra in 1950 [5], and by Nakayama

in 1951 [25], in various contexts. (Nakayama seems especially close
to the major concern of this book when he says '. . . duality systems
are defined purely in terms of representations . . . without appealing
to the group and its group operation.') Nonetheless, it is certainly
fair to say that the Hochschild–Mostow papers carry through the
Chevalley formulation of Tannaka duality in terms of $R(G)$ to a
complete theory that yields the structure of $R(G)$ for any analytic
group G, explains why there is a duality in the compact case
(Tannaka) and semisimple case (Harish–Chandra) as well as much
more.

We make no attempt here to summarize the Hochschild–Mostow
papers. Indeed, much of this book can be regarded as an exposition
of those papers. There are, however, two points we shall note now:
one is the discovery in [10] that if the analytic group G has a
faithful representation then G has a structure of algebraic variety
such that multiplications (on one side) by group elements are
morphisms. The coordinate ring of such a structure sits inside $R(G)$
and yields a good description of $R(G)$ (see Chapter 4). The other
point worth noting is that the last paper in the Hochschild–Mostow
series [16] deals essentially with the question of determining G
from $\mathrm{Mod}(G)$: given $R(G)$, as a Hopf algebra, how closely is G
known?

Using $R(G)$ as the dual object to G is, as we noted above,
somewhat disingenuous, if we want to regard G as unknown. A
mechanism to reinsert $\mathrm{Mod}(G)$ (as a category only) into the theory
appeared in Grothendieck's 1970 paper [4]. That paper is about
detecting whether a homomorphism of discrete groups of finite
type which induces an isomorphism on profinite completions is an
isomorphism, not about analytic groups, but the main technical
device used is to identify the profinite completion directly from
the category of modules for the group, as the group of 'tensor
automorphisms of the forgetful functor' (see (1.7) for a precise
definition).

This idea was then developed in full generality in Saavedra-
Rivano's 1972 lecture notes [27] on Tannakian categories: this is
a category of vector spaces with tensor product, meeting a few
technical conditions, which then turns out to be the category of

all modules for its group of tensor automorphisms (natural equivalences of the forgetful functor to vector spaces preserving tensor product), provided that 'module' is taken in the appropriate sense. Here, that means regarding the group of tensor automorphisms as an affine group scheme so modules are group-scheme modules.

It is interesting to note that Tannaka's original ideas were quite close to Grothendieck's formulation. His dual object $\hat{G}$ consisted only of irreducible representations of G (this is permitted since the compactness of G implies that all representations are completely reducible) and a representation of $\hat{G}$ in his sense is a selection of a linear automorphism of each element of $\hat{G}$ compatible with tensor product and direct sum. Because of the irreducibility of the elements of $\hat{G}$ and Schur's lemma, this selection is automatically compatible with G-module homomorphisms, and extends to give exactly a tensor automorphism of $\text{Mod}(G)$. The main point to keep in mind is that semisimplicity allows Tannaka to deal just with objects of $\text{Mod}(G)$; the homomorphisms are essentially just given by scalars so are carried along for free.

$\text{Mod}(G)$, when G is an analytic group, turns out to satisfy the axioms for a Tannakian category. So there is a group scheme such that $\text{Mod}(G)$ is the category of modules for it. An affine group scheme is completely determined by its Hopf algebra of global functions, and it is not terribly surprising that the group-scheme for $\text{Mod}(G)$ has $R(G)$ as its Hopf algebra. This was pointed out by Lubotzky in his 1978 PhD thesis from Bar-Ilan University.

It seems we are back to studying $R(G)$ again, but the difference is an important one: now $R(G)$ is constructed from $\text{Mod}(G)$ without direct reference to G (we have even used the notation $R(\text{Mod}(G))$ at some points to emphasize the relation). And with the benefit of hindsight, we can see how to proceed: we start with $\text{Mod}(G)$ and its group of tensor automorphisms, then construct $R(G)$ from that, and then go into the detailed study of $R(G)$. This in fact is the plan here, and we now turn to a detailed summary of the book.

3

This chapter-by-chapter summary is intended to let the potential reader know what is in store in a general way, as well as to let him know what the author considers to be the dominant themes of the work and how they are developed. Readers in search of specific results should consult the summaries which conclude each chapter.

Chapter 1 is entitled 'Definitions and Examples'. The definitions are those of representation and module for an analytic group G, and for $\text{Mod}(G)$, its group of tensor automorphisms $\text{Aut}_\otimes(\text{Mod}(G))$, and a canonical group homomorphism $\sigma: G \to \text{Aut}_\otimes(\text{Mod}(G))$. The examples (there are five labeled A to E) calculate $\text{Mod}(G)$, and as far as possible $\text{Aut}_\otimes(\text{Mod}(G))$, from first principles, for $G = \text{GL}_1\mathbb{C}$ (multiplicative group of complex numbers), $G = \mathbb{C}$ (additive group of complex numbers), $G = \mathbb{C} \rtimes \mathbb{C}^*$ (semidirect product of multiplicative group of complexes $\mathbb{C}^*$ acting on $\mathbb{C}$), $G = \mathbb{C} \rtimes \mathbb{C}$ (semidirect product of $\mathbb{C}$ acting on $\mathbb{C}$ via exponential $\mathbb{C} \to \text{GL}_1\mathbb{C}$), and $G = \mathbb{C}^{(2)} \rtimes \mathbb{C}$ (semidirect product of $\mathbb{C}$ acting on $\mathbb{C}^{(2)}$ via

$$\mathbb{C} \to \text{GL}_2\mathbb{C}, \, t \mapsto \begin{bmatrix} e^{\alpha t} & 0 \\ 0 & e^{\beta t} \end{bmatrix},$$

$\alpha, \beta \in \mathbb{C}$).

These groups are all elementary in the sense that they are at worst two-stage solvable. Determining their modules then is quickly reduced to finding modules for $\mathbb{C}^*$ or $\mathbb{C}$, which are the first two examples, subject to some compatibility rules. Now $\mathbb{C}^*$-modules are the same as graded vector spaces, and $\mathbb{C}$-modules are the same as vector spaces with a designated endomorphism, so finding the modules can be accomplished. Since the tensor product structure of $\text{Mod}(G)$ is central to our discussions, we also search for what might be called generators of $\text{Mod}(G)$ as a tensored category, and find them. Tensor automorphisms are determined by their values on these generators, and this gives us a handle on $\text{Aut}_\otimes(\text{Mod}(G))$.

The examples are elementary, but they point out most of the possibilities that occur in general. Here are the answers: (A) for $G = \mathrm{GL}_1\mathbb{C}$, σ is an isomorphism; (B) for $G = \mathbb{C}$, σ is one–one and $\mathrm{Aut}_\otimes(\mathrm{Mod}(G)) = \sigma(G) \times \mathrm{Hom}_{\mathrm{grp}}(\mathbb{C}, \mathbb{C}^*)$ (the second factor is the abstract group homomorphisms from $\mathbb{C}$ to $\mathbb{C}^*$; (C) for $G = \mathbb{C} \rtimes \mathbb{C}^*$, σ is again an isomorphism; (D) for $G = \mathbb{C} \rtimes \mathbb{C}$, σ is one–one and $\mathrm{Aut}_\otimes(\mathrm{Mod}(G)) = \sigma(G) \rtimes \mathrm{Hom}_{\mathrm{grp}}(\mathbb{C}, \mathbb{C}^*)$. (Example E is discussed below separately.)

Notice that it is for the vector group $\mathbb{C}$ (Example B) where the group of tensor automorphisms exceeds G, and the excess, it turns out, comes from the additive character group $\mathrm{Hom}(G, \mathbb{C})$ of analytic homomorphisms from G to $\mathbb{C}$; $\mathrm{Hom}(G, \mathbb{C})$ is the '$\mathbb{C}$' of the second factor of $\mathrm{Aut}_\otimes$. Example C has a large vector subgroup, but $\mathrm{Hom}(G, \mathbb{C}) = 1$ here and σ is an isomorphism. In Example D again, $\mathrm{Aut}_\otimes$ exceeds G, and the excess is again measured by the additive character group. In Examples A, B, and C, G can also be regarded as an algebraic group and these are the cases where $\sigma(G)$ is a direct factor of $\mathrm{Aut}_\otimes(\mathrm{Mod}(G))$. In general, one sees later that the excess (if any) of $\mathrm{Aut}_\otimes$ over $\sigma(G)$ is always measured by additive characters, and that of $\sigma(G)$ is a direct factor in $\mathrm{Aut}_\otimes$ exactly when G is algebraic.

The fifth example (actually it is a two parameter family of examples) is used somewhat differently. If the pair of parameters are algebraically the same in two of the groups (in the sense that they span isomorphic rational vector spaces), then a functor can be exhibited showing that the categories of modules for the two groups are equivalent. The parameters must be much closer for the groups to be isomorphic, so we produce lets of examples of nonisomorphic groups with the same module categories. This warns us, of course, to focus on $\mathrm{Mod}(G)$, not G, in our constructions later.

Chapter 2, 'Representative Functions', deals with the formal description of the category $\mathrm{Mod}(G)$ of finite-dimensional modules for the analytic group G. The representative functions on G coming from the module V are the functions on G, $g \mapsto f(gv)$ (for fixed $v \in V$ and $f \in V^*$), and $R(G)$ denotes the set of all such as f, v, and V vary over $\mathrm{Mod}(G)$. Now $R(G)$ also admits an intrinsic description as the set of holomorphic functions on G whose G-

translates span a finite-dimensional vector space. Under pointwise addition and multiplication of functions $R(G)$ is a complex algebra, in fact an integral domain.

It has further structure: using the group operators of G, $R(G)$ becomes a Hopf algebra. For example, if $f \in R(G)$ then the comultiplication sends f to $\sum h_i \otimes k_i$ if $f(xy) = \sum h_i(x)k_i(y)$ for all x and y in G. Both the algebra and coalgebra structure on $R(G)$ are defined using G. (This will be remedied later.) Moreover, there is a natural equivalence between the category $\text{Mod}(G)$ and the category of finite-dimensional comodules for $R(G)$. Thus, knowledge of $R(G)$, as a Hopf algebra, determines $\text{Mod}(G)$.

The converse of this assertion also holds: the Hopf algebra $R(G)$ can be produced directly from the category $\text{Mod}(G)$. This requires describing $R(G)$ other than as an algebra of functions on G, of course. It turns out that this can be done by describing $R(G)$ as an algebra of functions on $\text{Aut}_\otimes(\text{Mod}(G))$. This is accomplished via several intermediate steps.

Associated to the Hopf algebra $R(G)$ are two other groups, $\mathcal{G}(G)$, which is the group of $\mathbb{C}$-algebra homomorphisms from $R(G)$ to $\mathbb{C}$, and $\text{Propaut}(R(G))$, which is the group of $\mathbb{C}$-algebra automorphisms of $R(G)$ commuting with right G-translations. Both groups are isomorphic. Moreover, there are homomorphisms $\text{Aut}_\otimes(\text{Mod}(G)) \to \text{Propaut}(R(G))$ and $\mathcal{G}(G) \to \text{Aut}_\otimes(\text{Mod}(G))$ which are compatible with this isomorphism, so that all three groups are isomorphic. Finally, one shows that every G-module is naturally an $\text{Aut}_\otimes$-module and that $R(G)$ is the algebra of all functions on $\text{Aut}_\otimes(\text{Mod}(G))$ of the form $a \mapsto f(av)$ where $v \in V$ and $f \in V^*$ for some G-module V. Thus, we conclude that $R(G)$ can be produced, as a Hopf algebra, from $\text{Mod}(G)$. Moreover, we end up with a handle on the structure of $\text{Aut}_\otimes(\text{Mod}(G))$: we can think of it as the group $\mathcal{G}(G)$, so to determine $\text{Aut}_\otimes$ we can confine our attention to the Hopf algebra $R(G)$, which is a single (large) object on which G acts, rather than have to deal with the entire category $\text{Mod}(G)$. Now $\mathcal{G}(G)$ also receives a map from G, and we can regard $G \to \mathcal{G}(G)$ as a kind of 'universal representation' of G, in a sense which will be made clear below. Once we start dealing with this map, to be sure, we have moved our point of view in two senses: first, we are dealing directly with G again – this is not a map determined from the category Mod(G) – and second, we have shifted from modules to representations. In the module point of view the categorical properties of the set of all representations became tractible, but the group structure of G is in the background. Now, in the representation point of view, it comes to the front.

If $\rho : G \to GL(V)$ is a representation and $\bar{G}$ is the Zariski-closure of $\rho(G)$ in the algebraic group $GL(V)$, then $\rho(G)$ becomes a Zariski-dense analytic subgroup of the algebraic group $\bar{G}$. This is the situation studied in Chapter 3, 'Analytic subgroups of algebraic groups'. The main point of the chapter is to see that such subgroups are pretty close to being algebraic themselves.

An analytic subgroup of an algebraic group is an analytic group with a faithful representation. It turns out that the analytic groups G with a faithful representation are the groups which have a nucleus: that is, a connected, simply-connected solvable normal subgroup K such that G/K is reductive (reductive means faithfully representable and all modules semisimple). This is actually one of the last results of the chapter; it begins by considering groups with nuclei. If G is one such, and K is a nucleus, then there is a reductive subgroup P of G with $G = KP$ and $K \cap P = \{e\}$. Next, it is shown that if G has such a decomposition $G = KP$ then G can be embedded in an algebraic group $\bar{G}$ via an analytic embedding f such that $f(G)$ is Zariski-dense in $\bar{G}$, and there is a torus T in $\bar{G}$ such that $\bar{G} = f(G)T$, $T \cap f(G) = e$, and T centralizes $f(P)$. We call the triple $(\bar{G}, f, T)$ a split hull of G.

We pause a moment to reflect on the importance of the torus T. Suppose we have embedded G in $GL(V)$ and chosen a basis so that the solvable subgroup K is in upper triangular form. The intersection of K with the diagonal maximal torus of $GL(V)$ is an analytic, but possibly not algebraic, subgroup and in taking its Zariski-closure T appears. If we look at the diagonal entries as functions on K, they have the form $x \mapsto e^{a(x)}$ where $a : K \to \mathbb{C}$ is an analytic homomorphism. It is the presence of such functions that separates the analytic theory from the algebraic theory, and the torus T is a group-theoretical record that there are such analytic but not algebraic characters present.

These split hulls have a strong permanence property: if $(\bar{G}, f, T)$ is a split hull of G and if H is an algebraic group with $p : H \to \bar{G}$ a morphism and $g : G \to H$ an analytic homomorphism with $pg = f$ and $g(G)$ Zariski-dense in H, then $p^{-1}(T)$ is a torus in H and $(H, g, p^{-1}(T))$ is also a split hull of G.

In order to appreciate these results, we need to think a little about how analytic homomorphisms from G to an algebraic group $\bar{G}$ are connected to representations. The algebraic group $\bar{G}$ can be regarded as a subgroup of $\mathrm{GL}(V)$ for some vector space V. Moreover, a polynomial function on $\bar{G}$ pulls back to a representative function on G. If the image of G is Zariski-dense in $\bar{G}$ this defines an injective homomorphism from $\mathbb{C}[\bar{G}]$, the coordinate ring of $\bar{G}$, to $R(G)$. Conversely, if A is any finitely-generated Hopf subalgebra of $R(G)$, A can be regarded as the coordinate ring of some algebraic group $\bar{G}$, and there is an analytic homomorphism $G \to \bar{G}$ with Zariski-dense image such that $\mathbb{C}[\bar{G}]$ embeds in $R(G)$ with image A. We can choose A large enough to contain any finite subset of $R(G)$. Combined with the results of the preceding paragraph, this implies that every representative function can be regarded as coming from a G-module V with the Zariski-closure of G in $\mathrm{GL}(V)$ being a split hull of G.

This point of view is exploited in Chapter 4, 'Structure of algebras of representative functions'. In general terms, the plan is the following: think of $R(G)$ as the direct limit of finitely-generated Hopf subalgebras, say $R(G) = \mathrm{dir\ lim}\ A_i$, where A_i is to be thought of as the coordinate ring of the algebraic group $\bar{G}_i$. If we assume the inverse system of groups $\bar{G}_i$ starts with a split hull of G, then the permanence theorem of Chapter 3 allows us to choose a torus in each G_i so that we have an inverse system of split hulls of G. If we pass to the inverse limit, we obtain the group $\mathscr{G}(G)$ of Chapter 2, and a subgroup T of it which is an inverse limit of tori, such that $\mathscr{G}(G)$ is the semidirect product of the image of G and T.

We recall that $\mathscr{G}(G)$ is the same as $\mathrm{Aut}_\otimes(\mathrm{Mod}(G))$ and is our basic object of study. The difference between G and $\mathscr{G}(G)$ is T, so we need to know what T is and where it comes from.

Thus, suppose $(\bar{G}_i, f_i, T_i)$ is one of the split hulls in the above system. Then there is an algebraic group homomorphism d_i from

$\bar{G}_i$ to T_i, which is the identity on T_i, such that the compositions of the characters of T_i with d_i are precisely those characters of G in $\mathbb{C}[\bar{G}_i]$ which factor through $\mathbb{C}$, the 'exponential additive' characters. Further, the d_is are compatible in the inverse system.

Two consequences follow: first, $\mathbb{C}[\bar{G}_i]$ is the group algebra of the group Q_i of exponential additive characters in $\mathbb{C}[\bar{G}_i]$, with coefficients in the coordinate ring D_i of the homogeneous space $\bar{G}_i/T_i$ so that in the limit $R(G) = \text{dir lim}(D_i[Q_i])$, and second, T_i has character group Q_i so that as an abstract group $T_i = \text{Hom}_{\text{grp}}(Q_i, \mathbb{C}^*)$ and in the limit $T = \text{Hom}_{\text{grp}}(Q, \mathbb{C}^*)$, where $Q = \text{dir lim } Q_i$ is the group of all exponential additive characters of G. Further, the choice of the T_i in the inverse system guarantees that all the D_i are equal to the first one (call it D) so that $R(G)$ is the group algebra $D[\mathbf{Q}]$ of Q with coefficients in D.

Thus, the exact measure of the difference between G and $\text{Aut}_{\otimes}(\text{Mod}(G))$ is $Q = \exp(\text{Hom}(G, \mathbb{C}))$, where the homomorphisms are of analytic groups.

The group algebra description $R(G) = D[Q]$ only gives the algebra structure of $R(G)$. Nonetheless, by its construction we also know some more of the properties of D: it is finitely-generated as a $\mathbb{C}$-algebra, it is stable under right translations by elements of G, and (regarded as an algebra of functions on G) the maximal ideals of D are in one–one correspondence with the elements of G. This means we can use D to regard G as an algebraic variety with coordinate ring D, and if we do so then the left translations on G by elements of G are all morphisms. So we call D a left algebraic group structure on G. Now D came from a split hull of G and, in fact, any subalgebra of $R(G)$ with the above three properties must arise from a split hull of G. In Chapter 3 we prove the existence of split hulls in which the torus commuted with the reductive subgroup. The left algebraic group structures coming from such split hulls are called basal.

To understand fully the decompositions $R(G) = D[Q]$, we need to know what the Ds are like and how to find them. This can be done, at least in the basal case, and is the subject of Chapter 5, 'Left algebraic groups'. It is best described by assuming we start with a basal left algebraic group structure D on G, but the description works equally well in reverse and yields the desired classification of left algebraic group structure.

To D is associated a split hull $(\bar{G}, f, T)$ and a nucleus K. The centralizer C of T in G (really in $f(G)$) is called the core of D (it has a direct description in terms of D also) and it does turn out to hold the essence of D, as will be explained shortly.

Now D, being an algebra of functions on G, restricts to an algebra of functions on K denoted $D|K$. Also, $D|K$ is again a left algebraic group structure on K and its core is $C \cap K$ (this statement about cores uses the fact that D is basal); K itself is an analytic group and is simply-connected and solvable. We find that left algebraic group structures on such groups are completely determined by their cores, and that their cores are Cartan subgroups. We can even show how to write down all the left algebraic group structures explicitly in this case.

It also turns out that once we know $C \cap K$, just as a group, we know enough to reproduce D: if R is the radical of G, then $K = (R, G)(C \cap K)$ and C is the centralizer of $C \cap K$ in G. From this data we find D. So in the end, the Cartan subgroup $C \cap K$ of K determines D. And, as we remarked above, this is all reversible.

Finally, since the nucleus K is simply-connected we might as well deal with Cartan subalgebras of its Lie algebra. The chapter concludes with a theorem which allows us to recognize which subalgebras of the Lie algebra of G are Lie algebras of nuclei (the criteria are simple: the subalgebra must be an ideal and a direct sum complement to the Lie algebra of a maximal reductive subgroup of G) and, hence, we can, in principle, construct the left algebraic group structures on G completely from its Lie algebra.

4

We assume the reader of this volume is familiar with the basics of complex Lie group theory and characteristic zero affine algebraic group theory. We also use some parts of the theory of Hopf algebras and comodules; although not much previous knowledge of these subjects is assumed, interested readers can consult [28] for details and more information.

We use, almost from the beginning, the basic facts about commutative analytic groups and their homomorphisms to one-dimensional groups. This material is not always included in the basic texts, so for convenience we have recorded the necessary results in an appendix.

Terminology is, as far as possible, standard, and in any event is explained in the 'Notation and conventions' section which precedes this introduction. The section headed 'Notes' cites the sources of the main concepts and results presented here. The section headed 'Bibliography' is intented also as a guide to further reading, as well as a list of works cited in the 'Introduction' and 'Notes'.

1
Definitions and examples

In this chapter, we define the category of representations of an analytic group and study several examples from an elementary viewpoint.

Definition 1.1 Let G be an analytic group and V a (finite-dimensional) complex vector space. A *representation* of G on the space V is an analytic homomorphism $\rho : G \to \mathrm{GL}(V)$.

If ρ is a representation of G on V, we can, as usual, define an action of G on V by the formula $g(v) = \rho(g)(v)$ for g in G and v in V. This action satisfies the following conditions.

(1.2a)
$$\begin{cases} g(\alpha v + v') = \alpha g(v) + g(v') \\ (gg')(v) = g(g'v) \\ e(v) = v \end{cases}$$

(1.2b) for all g, g' in G, v, v' in V and α in $\mathbb{C}$, and $g \mapsto f(g(v))$ is an analytic function on G for all v in V and f in V^*.

Conversely, suppose G acts on a vector space V such that conditions (1.2) are satisfied. For each $g \in G$, define $\rho(g) : V \to V$ by $\rho(g)(v) = g(v)$. Then (1.2a) shows that $\rho : G \to \mathrm{GL}(V)$ is a well-defined group homomorphism, and (1.2b) shows that ρ is analytic.

Definition 1.3 Let G be an analytic group. A *G-module* is a (finite-dimensional) complex vector space with a G-action satisfying conditions (1.2).

The notions of 'representation' and 'module' are completely equivalent. For what might be termed structural investigations, representations are the most convenient to deal with (see Chapter

15

3), while it is much easier to express categorical notions in module language. This applies notably to the definition of homomorphism.

Definition 1.4   Let G be an analytic group, V and W G-modules. A *G-module homomorphism* $T\colon V \to W$ is a complex linear transformation such that $T(g(v)) = g(T(v))$ for all g in G, v in V.

It is a trivial consequence of the definition that the identity linear transformation on a G-module is a G-module homomorphism, and that the composition of two G-module homomorphisms is also a G-module homomorphism. Thus there is a category of G-modules.

Notation 1.5   Let G be an analytic group. $\mathrm{Mod}(G)$ denotes the category of G-modules and G-module homomorphisms.

In $\mathrm{Mod}(G)$ we have submodules and quotient modules, in the obvious senses, and kernels (respectively, cokernels) of G-module homomorphisms are submodules (respectively, quotient modules). Moreover, we have the following additional constructions:

If V is a G-module, V^* becomes a G-module under the formula $g(f)(v) = f(g^{-1}(v))$ for $f \in V^*$, $g \in G$, $v \in V$.

If V and W are G-modules, $V \oplus W$ becomes a G-module under the formula $g((v, w)) = (g(v), g(w))$ for $g \in G$ and $(v, w) \in V \oplus W$.

If V and W are G-modules, $V \otimes_{\mathbb{C}} W$ becomes a G-module under the formula $g(\sum v_i \otimes w_i) = \sum g(v_i) \otimes g(w_i)$ for $g \in G$, $\sum v_i \otimes w_i \in V \otimes_{\mathbb{C}} W$.

We summarize these properties in technical language for later reference.

Theorem 1.6.   Let G be an analytic group. Then $\mathrm{Mod}(G)$ is an abelian subcategory of the category $\mathrm{Mod}(\mathbb{C})$ of (finite-dimensional) complex vector spaces, closed under linear dual and tensor product.

The major question this work investigates is 'To what extent does $\mathrm{Mod}(G)$, as a category, determine the analytic group G?' The tensor product operation in $\mathrm{Mod}(G)$ is a major tool in this study. It enters via the concept, due to Grothendieck, of a tensor automorphism of the forgetful functor $\mathrm{Mod}(G) \to \mathrm{Mod}(\mathbb{C})$. Put concretely, this means the following.

Definition 1.7 Let G be an analytic group. A tensor automorphism of the forgetful functor $\text{Mod}(G) \to \text{Mod}(\mathbb{C})$ is a set of $\mathbb{C}$-linear automorphisms $a_V : V \to V$ for each V in $\text{Mod}(G)$ such that:
   (1) If $f : V \to W$ is in $\text{Mod}(G)$, $a_W f = f a_V$ in $\text{Mod}(\mathbb{C})$;
   (2) $a_{V \otimes W} = a_V \otimes a_W$ for V, W in $\text{Mod}(G)$.

It is clear that if $\{a_V\}$ and $\{b_V\}$ are two such tensor automorphisms, then so is their composite $\{c_V = a_V b_V\}$; also $\{d_V = a_V^{-1}\}$ is a tensor automorphism, and so is $\{id_V\}$. Thus, the set of all tensor automorphisms is a group.

Notation 1.8 Let G be an analytic group. $\text{Aut}_\otimes(\text{Mod}(G))$ denotes the group of tensor automorphisms of the forgetful functor $\text{Mod}(G) \to \text{Mod}(\mathbb{C})$.

We can always produce tensor automorphisms by the following device: let $g \in G$ and let V be a G-module. Define $\sigma(g)_V : V \to V$ by $\sigma(g)_V(x) = g(x)$. Then $\sigma(g)_V$ is a $\mathbb{C}$-linear automorphism of V (with inverse $\sigma(g_V^{-1})$) and it is immediate from the definitions that $\{\sigma(g)_V\}$ is a tensor automorphism. We record this construction for later reference.

Proposition 1.9 Let G be an analytic group. There is a group homomorphism $\sigma : G \to \text{Aut}_\otimes(\text{Mod}(G))$ given by $\sigma(g) = \{\sigma(g)_V\}$, where $\sigma(g)_V(x) = g(x)$.

In general terms, we should regard the group of tensor automorphisms as the closest we can come to G from $\text{Mod}(G)$ (this will be made more precise in Chapter 2), and the position of G inside $\text{Aut}_\otimes(\text{Mod}(G))$ via the homomorphism σ of (1.9) as a measure of how much we miss.

The remainder of this chapter is devoted to the study of several elementary examples, where we can compute $\text{Mod}(G)$, and $\text{Aut}_\otimes(\text{Mod}(G))$ explicitly enough to see how the latter relates to G via σ. These examples use freely Lie group and Lie algebra methods, often without notice. The reader is encouraged to go through them carefully: a large part of the general theory to be developed here is already exhibited in these examples.

Example A $G = GL_1\mathbb{C}$, the multiplicative group of complex numbers.

In this case, if $V \in \text{Mod}(G)$, and $i \in Z$, let $V_i = \{x \in V \mid g(x) = g^i x$ for all $g \in G\}$. (In the definition of V_i, the equation $g(x) = g^i x$ refers to g in G acting on x in the G-module V on the left, and to g^i in $\mathbb{C} - \{0\}$ acting as a scalar on x in the vector space V on the right.)

Then V_i is also in $\text{Mod}(G)$, and $V = \bigoplus_{i \in Z} V_i$. (Since V is finite-dimensional, most V_i are zero.)

Conversely, given any set $\{V_i \mid i \in Z\}$ of finite-dimensional vector spaces, almost all zero, we can make their sum $V = \bigoplus_{i \in Z} V_i$ into a G-module by the formula $g(\sum x_i) = \sum g^i x_i$ (where $x_i \in V_i$).

Finally, if $f: V \to W$ is any morphism in $\text{Mod}(G)$, then $f(V_i) \subseteq W_i$ for all $i \in Z$.

It follows from the above considerations that $\text{Mod}(G)$ is isomorphic to the category of finite-dimensional Z-graded vector spaces. Further, the tensor product in $\text{Mod}(G)$ becomes the graded tensor product: if $V = \bigoplus V_i$ and $W = \bigoplus W_i$ then $(V \otimes W)_k = \bigoplus_{i+j=k} (V_i \otimes W_j)$.

Let $\mathbb{C}[n]$, for $n \in Z$, be the graded vector space with $(\mathbb{C}[n])_i = 0$ if $i \neq n$, and $\mathbb{C}$ if $i = n$. If $V = \bigoplus V_i$ is a graded vector space with V_i of dimension n_i, then $V = \bigoplus (\mathbb{C}[i])^{(n_i)}$. Also, $\mathbb{C}[n] \otimes \mathbb{C}[m] = \mathbb{C}[n+m]$, and $(\mathbb{C}[n])^* = \mathbb{C}[-n]$. Thus every graded vector space can be produced, from duals, direct sums, and tensor products, from $\mathbb{C}[0]$ and $\mathbb{C}[1]$.

We want to determine $\text{Aut}_\otimes(\text{Mod}(G))$. Let α be a tensor automorphism. It follows from general considerations that $\alpha_{V \oplus W} = \alpha_V \oplus \alpha_W$ and $\alpha_{V^*} = (\alpha_V^*)^{-1}$. Thus, thinking of $\text{Mod}(G)$ as graded vector spaces, α will be determined by its values, α_0 and α_1, on $\mathbb{C}[0]$ and $\mathbb{C}[1]$, respectively. Since both of these vector spaces are $\mathbb{C}$ in their one nontrivial place, we can think of α_0 and α_1 as nonzero scalars. The relation $\mathbb{C}[0] \otimes \mathbb{C}[0] = \mathbb{C}[0]$ forces $\alpha_0^2 = \alpha_0$, or $\alpha_0 = 1$. Thus α is completely determined by α_1. If we think of α_1 as an element of G, then $\sigma(\alpha_1)$, as defined in (1.9), is such that $\sigma(\alpha_1)_1 = \alpha_1$.

We have shown that $\sigma: G \to \text{Aut}_\otimes(\text{Mod}(G))$ is onto. If g is in the kernel, then $\sigma(g)_1 = 1$, and the definition of $\sigma(g)$ shows that $g = 1$, so σ is an isomorphism.

Example B $G = (\mathbb{C}, +)$, the additive group of complex numbers. Let V be in Mod(G). It follows from Lie algebra considerations, and the fact that G is simply connected, that there is a unique linear endomorphism A of V such that $g(v) = e^{gA}(v)$ for all v in V. Thus the objects of Mod(G) are pairs (V, A), where $A \in$ End(V), and every such pair gives rise to a G-module via the exponential of A. Further, if $f: (V, A) \to (W, B)$ is a morphism in Mod(G), $Bf = fA$. We can think of a pair (V, A) as a module over the polynomial ring $\mathbb{C}[X]$ via the action $Xv = A(v)$ for $v \in V$, and then Mod(G) becomes the category of finite-dimensional modules over $\mathbb{C}[X]$. (All we have really done here is observe that G-modules are the same as Lie(G)-modules, and the universal enveloping algebra of Lie(G) is $\mathbb{C}[X]$.) If V and W are finite-dimensional $\mathbb{C}[X]$-modules, then $V \otimes W$ is a $\mathbb{C}[X]$-module via $X(v \otimes w) = (Xv \otimes w) + (v \otimes Xw)$, and it is easy to check that this action is the tensor product action in Mod(G).

Because of the above, we regard Mod(G) as the category of finite-dimensional $\mathbb{C}[X]$-modules. Now $\mathbb{C}[X]$ is a principal ideal domain and a finite-dimensional $\mathbb{C}[X]$-module is just a finitely-generated torsion module. From the usual structure theory for such modules, we know that they are all direct sums of modules of the form $\mathbb{C}[X]/(X - a)^n$, for $a \in \mathbb{C}$ and n a positive integer.

For convenience, we establish the following notation for these modules: $V_n(a) = \mathbb{C}[X]/(X - a)^n$, $V_n = V_n(0)$, and $V(a) = V_1(a)$. Note that $V_n(a)$ is n-dimensional with basis $1, X - a, \ldots,$ $(X - a)^{n-1}$. In this basis, $X(X - a)^i = a(X - a)^i + (X - a)^{i+1}$ is the $\mathbb{C}[X]$ action. Now consider the module $V(a) \otimes V_n$: this is n-dimensional with basis $1 \otimes 1, 1 \otimes X, \ldots, 1 \otimes X^{n-1}$, and in the basis $X(1 \otimes X^i) = X(1) \otimes X^i + 1 \otimes X^{i+1} = a(1 \otimes X^i) + (1 \otimes X^{i+1})$ is the $\mathbb{C}[X]$ action. Thus $V(a) \otimes V_n$ and $V_n(a)$ are isomorphic via the map which, on the chosen bases, sends $1 \otimes X^i$ to $(X - a)^i$. To determine tensor automorphisms, therefore, of Mod(G), we can look at the subcategory of $V(a)$s and the subcategory of V_ns separately.

To deal with the $V(a)$ case, we note that $V(a) \otimes V(b)$ is isomorphic to $V(a + b)$: if $1 \otimes 1$ is the basis of $V(a) \otimes V(b)$, $X(1 \otimes 1) = X(1) \otimes 1 + 1 \otimes X(1) = a \otimes 1 + 1 \otimes b = (a + b)(1 \otimes 1)$. If α is a tensor automorphism of $\text{Mod}(G)$, let α_a be α on $V(a)$. Since $V(a)$ is one-dimensional, we can regard α_a as an element of $\mathbb{C}^+$. Then the isomorphism $V(a) \otimes V(b) \to V(a+b)$ shows that $\alpha_a \alpha_b = \alpha_{a+b}$, or that $a \mapsto \alpha_a$ is in $\text{Hom}(\mathbb{C}, \mathbb{C}^*)$.

To deal with the V_n case, we have to consider some additional mappings. We consider the basis $y_i = X^{n-i-1}$, $i = 0, \ldots, n-1$ of V_n, and we let $y_{-1} = 0$. Then in this basis the $\mathbb{C}[X]$ action is $X(y_i) = y_{i-1}$ for $i = 0, \ldots, n-1$. Finally, we let $z^i = i! y_i$, $i = 0, \ldots, n-1$. This is also a basis of V_n, and $X(z^i) = i! y_{i-1} = iz^{i-1}$. Finally, we let $p_{n,m}: V_n \otimes V_m \to V_{n+m-1}$ be given by $p_{n,m}(z^i \otimes z^j) = z^{i+j}$. We verify that this is a $\mathbb{C}[X]$-homomorphism: $X(z^i \otimes z^j) = iz^{i-1} \otimes z^j + z^i \otimes jz^{j-1}$, so $p_{n,m}(X(z^i \otimes z^j)) = iz^{i+j-1} + jz^{i+j-1} = (i+j)z^{i+j-1} = X(z^{i+j}) = X(p_{n,m}(z^i \otimes z^j))$. It is also clear that $p_{n,m}$ is onto.

Now let α be a tensor automorphism of $\text{Mod}(G)$ and let α_n be α on V_n. Then, $(\alpha_{n+m-1})(p_{n,m}) = (p_{n,m})(\alpha_n \otimes \alpha_m)$ for all n, m. For $n = m = 1$, we conclude that $\alpha_1 = 1$, and for $m = 2$, n arbitrary, we have $\alpha_{n+1}(p_{n,2}) = p_{n,2}(\alpha_n \otimes \alpha_2)$. Since $p_{n,2}$ is onto, this means that α_2 and α_n determine α_{n+1}, and, hence, that α_2 determines all α_n.

To study α_2, we use the basis y_0, y_1 of V_2 described above. There is an exact sequence $V_1 \to V_2 \to V_1$ of $\mathbb{C}[X]$-modules, where the first map sends the generator 1 of V_1 to y_0 and the second map sends y_1 to 1 and y_0 to 0. Thus, we have a commutative diagram:

$$
\begin{array}{ccccc}
V_1 & \longrightarrow & V_2 & \longrightarrow & V_1 \\
\alpha_1 \downarrow & & \alpha_2 \downarrow & & \alpha_1 \downarrow \\
V_1 & \longrightarrow & V_2 & \longrightarrow & V_1.
\end{array}
$$

The left-hand square shows that $\alpha_2(y_0) = y_0$ and the right-hand square shows that $\alpha_2(y_1) = y_1 + ay_0$ for some $a \in \mathbb{C}$. Let $\sigma(a)_2$ be the value of $\sigma(a)$ on V_2. Since $X(y_0) = 0$ and $X(y_1) = y_0$, the linear transformation A on V_2 corresponding to multiplication by X is such that $A^2 = 0$. Thus, $\sigma(a)_2 = e^{aA} = I + aA$, so $\sigma(a)_2(y_0) = y_0$ and $\sigma(a)_2(y_1) = y_1 + ay_0$ and $\sigma(a)_2 = \alpha_2$. (We also note that the equation $\sigma(a)_2(y_1) = y_1 + ay_0$ shows that σ is injective.)

We now assemble all the above discussion to determine $\text{Aut}_\otimes(\text{Mod}(G))$. Let α be a tensor automorphism of $\text{Mod}(G)$. We can find $a \in \mathbb{C}$ such that α and $\sigma(a)$ agree on all V_n, so that $\alpha' = \alpha\sigma(-a)$ is the identity on all V_n. Then, α' is determined by its values α_b' on the $V(b)$, i.e., by the homomorphism corresponding to α' in $\mathrm{Hom}(\mathbb{C}, \mathbb{C}^*)$. Conversely, given any f in $\mathrm{Hom}(\mathbb{C}, \mathbb{C}^*)$ we can define a tensor automorphism $\alpha(f)$ of $\mathrm{Mod}(G)$ whose value on $V(b) \otimes V_n$ is $f(b) \otimes id$ (we extend this in the obvious way to direct sums to get automorphisms on all of $\mathrm{Mod}(G)$). Further, it is clear that $\alpha(f)\sigma(a) = \sigma(a)\alpha(f)$ on $V(b) \otimes V_n$ for all $a \in \mathbb{C}$, since $V(b)$ is one-dimensional so $\alpha(f)$ and $\sigma(a)$ commute on it. We can define a map $\mathbb{C} \times \mathrm{Hom}(\mathbb{C}, \mathbb{C}^*) \to \mathrm{Aut}_\otimes(\mathrm{Mod}(G))$ by sending (a, f) to $\sigma(a)\alpha(f)$. This is a surjective group homomorphism, and if (a, f) is in the kernel, then $\sigma(a) = \alpha(f^{-1})$ so $\sigma(a)$ is the identity on V_2 and $a = 0$.

The group $\mathrm{Hom}(\mathbb{C}, \mathbb{C}^*)$ which appears in the determination of $\mathrm{Aut}_\otimes(\mathrm{Mod}(G))$ seems intractible. For later use, we note that $\mathrm{Hom}(\mathbb{C}, \mathbb{C}^*) = \mathrm{Alg}_\mathbb{C}(\mathbb{C}[\mathbb{C}], \mathbb{C})$. In other words, $\mathrm{Hom}(\mathbb{C}, \mathbb{C}^*)$ is the set of $\mathbb{C}$-rational points of the group scheme with Hopf algebra $\mathbb{C}[\mathbb{C}]$ (the complex group algebra of the discrete group $\mathbb{C}$). This is quite close to being an algebraic group, as we will see below.

Example C G is the semidirect product of $\mathbb{C}$ and $\mathbb{C}^*$, with $\mathbb{C}^*$ acting as $GL_1\mathbb{C}$: so $G = \{(a, t) \,|\, a \in \mathbb{C}, t \in \mathbb{C}^*\}$, and $(a, t)(b, s) = (a + tb, ts)$.

G contains subgroups, $\mathbb{C} \times \{1\}$ and $\{0\} \times \mathbb{C}^*$, isomorphic to $\mathbb{C}$ and $\mathbb{C}^*$, and we regard these isomorphisms as identities. Thus, elements of G can be uniquely written as at with $a \in \mathbb{C}$ and $t \in \mathbb{C}^*$, and $ta = t(a)t$, where $t(a)$ is the action of t (in $GL_1\mathbb{C}$) on a (in $\mathbb{C}^{(1)}$), or $tat^{-1} = t(a)$.

Now let V be in $\mathrm{Mod}(G)$. Then we can regard V as both a $\mathbb{C}$- and a $\mathbb{C}^*$-module also. By Example B, the $\mathbb{C}$-module structure is given by a linear transformation N of V such that $a(v) = e^{aN}(v)$ for all $a \in \mathbb{C}$ and $v \in V$. By Example A, the $\mathbb{C}^*$-module structure is given by a grading of V, where $V_i = \{v \in V \,|\, t(v) = t^i v$ for all $t \in \mathbb{C}^*\}$. We need to see how these two module structures are linked.

For $g \in G$, let $\rho(g)$ be the linear transformation on V given by $\rho(g)(v) = g(v)$. Then since $tat^{-1} = t(a)$, $\rho(t)\rho(a)\rho(t)^{-1} = \rho(t(a))$.

Now $\rho(a) = e^{aN}$ and $\rho(t(a)) = e^{taN}$, so $e^{taN} = \rho(t) e^{aN} \rho(t)^{-1} = \exp(a\rho(t)N\rho(t)^{-1})$. It follows that $\rho(t)N\rho(t)^{-1} = tN$, or $\rho(t)N = tN\rho(t)$. Now if $v \in V_i$, $t(N(v)) = \rho(t)N(v) = tN(t^i v) = t^{i+1}N(v)$, so $N(V_i) \subseteq V_{i+1}$. Thus, V is a graded vector space with a linear endomorphism of degree 1.

Conversely, suppose $W = \oplus W_i$ is a (finite-dimensional) graded vector space and N a linear endomorphism of W of degree 1; W is a $\mathbb{C}$ -module via $a(w) = e^{aN}(w)$ and w is a $\mathbb{C}^*$ -module via $t(w) = t^i(w)$ if $w \in W_i$. Let $w \in W_i$, and consider $(tat^{-1})(w) = t(a(t^{-i}w)) = t^{-i}(t(a(w)))$. Now $a(w) = e^{aN}(w) = \sum (k!)^{-1}(aN)^k(w)$, and $(aN)^k(w) \in W_{i+k}$. Thus, $t(a(w)) = \sum (k!)^{-1}t^{i+k}(aN)^k(w)$, and $t^{-i}(t(a(w))) = \sum (k!)(taN)^k(w) = (t(a))(w)$. Thus, $(tat^{-1})(w) = (t(a))(w)$, so W is in fact a G -module.

It is clear from Examples A and B that if V and W are G -modules and $f: V \to W$ a G -homomorphism, then $f(V_i) \subseteq W_i$ for each i and $fN = N'f$ where N, N' are the degree 1 endomorphisms of V and W .

Thus we can sum up by saying that $\text{Mod}(G)$ is the category of graded vector spaces with degree 1 endomorphisms. We will determine its tensor automorphisms now.

We want to consider some special objects in $\text{Mod}(G)$: $W_n[i]$, for $i \in Z$ and n nonnegative, is the n -dimensional vector space with basis $x_0, \ldots, x_{n-1}$, grading such that x_k has degree $k + i$, and degree 1 endomorphism $N(x_k) = x_{k+1}$ (where $x_n = 0$). These objects generate $\text{Mod}(G)$ in the following sense: If V is an object in $\text{Mod}(G)$, $v \in V_i$, and if $N^n(v) = 0$, where N is the degree 1 endomorphism of V , then the linear transformation $f: W_n[i] \to V$ given by $f(x_k) = N^k(v)$ is a morphism in $\text{Mod}(G)$ whose image contains v . Thus every object of $\text{Mod}(G)$ can, by repeating this process, be viewed as a homomorphic image of a direct sum of copies of $W_n[i]$ s, for various n and i . This in turn implies that tensor automorphisms of $\text{Mod}(G)$ are determined by their values on the $W_n[i]$.

The $W_n[i]$ are themselves described more simply: we first consider $W_1[i] \otimes W_n[0]$. This is a n -dimensional vector space with basis $x_0 \otimes x_0, \ldots, x_0 \otimes x_{n-1}$, where $x_0 \otimes x_k$ has degree $(0+i)+(k+0) = k+i$, and $N(x_0 \otimes x_k) = N(x_0) \otimes x_k + x_0 \otimes N(x_k) = x_0 \otimes x_{k+1}$. Thus, $W_1[i] \otimes W_n[0]$ is isomorphic to $W_n[i]$. Thus, tensor automorphisms of $\text{Mod}(G)$ are determined by their values on the $W_1[i]$ and the $W_n[0]$. For easy notation we write $W[i] = W_1[i]$ and $W_n = W_n[0]$.

Now it is easy to verify that $W[i] \otimes W[j] = W[i+j]$ and that $(W[i])^* = W[-i]$, so a tensor automorphism's value on $W[1]$ determines its value on all $W[i]$. There is a similar relation between W_2 and the other W_n. Let $W = W_2^{\otimes n}$, the n-fold tensor power of W_2, and consider $x^0 = x_0 \otimes \cdots \otimes x_0$ in W_0. We will need some additional notations: if $A \subseteq \{1, 2, \ldots, n\}$, let $x_A = x_{i_1} \otimes \cdots \otimes x_{i_n}$ where $i_k = 1$ if $k \in A$, 0 otherwise. Then, $x^0 = x_\varnothing$, and it is easy to check that $N(x_A) = \sum \{x_B \mid B \supseteq A \text{ and } |B| = |A| + 1\}$. Finally, let $x^i = \sum \{x_A \mid |A| = i\}$. Then $N(x^i) = \sum \{\sum x_B \mid B \supseteq A, |B| = |A| + 1, |A| = i\}$. Since every subset B of cardinality $i + 1$ has $i + 1$ subsets of cardinality i, each B occurs $i + 1$ times so $N(x^i) = (i+1)x^{i+1}$. Now let $y_i = i! x^i$ for $i = 0, \ldots, n - 1$. Then $N(y_i) = (i+1)! x^{i+1} = y_{i+1}$. We thus have an injection $W_n \to W_2^{\otimes n}$ given by $x_i \mapsto y_i$. Thus, the value of a tensor automorphism on W_n is determined by its value on W_2.

Now let α be a tensor automorphism of $\text{Mod}(G)$. To determine α, we need only know its values α_2 on W_2 and α_1 on $W[1]$. We remark that the value of α on $W[0] = W_1$ is determined from the isomorphism $W[0] \otimes W[0] = W[0]$; so it must be the identity.

Since $W[1]$ is one-dimensional, α_1 is a nonzero scalar t. To determine α_2, we consider the sequence $W[1] \to W_2 \to W_1$ where the first map sends x_0 to x_1 and the second sends x_0 to x_0 and x_1 to 0. These are morphisms in $\text{Mod}(G)$, and, as in Example B, the commutative diagram

$$
\begin{array}{ccccc}
W[1] & \longrightarrow & W_2 & \longrightarrow & W_1 \\
\downarrow {\scriptstyle \alpha_1} & & \downarrow {\scriptstyle \alpha_2} & & \downarrow {\scriptstyle 1} \\
W[1] & \longrightarrow & W_2 & \longrightarrow & W_1
\end{array}
$$

shows that $\alpha_2(x_1) = t x_1$ and $\alpha_2(x_0) = x_0 + a x_1$ for some $a \in \mathbb{C}$.

Now consider the element $g = at$ of G and the tensor automorphism $\sigma(g)$. The degree 1 endomorphism N of W_2 has $N^2 = 0$, so $e^{aN} = I + aN$. Thus, $g(x_0) = a(t(x_0)) = a(x_0) = x_0 + a x_1$ and $g(x_1) = a(t(x_1)) = a(t x_1) = t x_1$, so $\sigma(g)$ on W_2 is α_2. On $W[1]$, $g(x_0) = t x_0$, so $\sigma(g)$ on $W[1]$ is α_1. Since α_1 and α_2 determine α, $\sigma(g) = \alpha$.

It is clear from the calculation of $\sigma(g)$ on W_2 and $W[1]$ above that $\sigma(g)$ determines g, so $\sigma: G \to \mathrm{Aut}_\otimes(\mathrm{Mod}(G))$ is an isomorphism.

Example D G is the semidirect product of $\mathbb{C}$ with $\mathbb{C}$, where the latter acts on the former via $\exp: \mathbb{C} \to \mathrm{GL}_1\mathbb{C}$; so $G = \{(a, b) \mid a, b \in \mathbb{C}\}$, and $(a, b)(a', b') = (a + e^b a', b + b')$.

 G contains subgroups $\mathbb{C} \times \{0\}$ and $\{0\} \times \mathbb{C}$, isomorphic to $\mathbb{C}$, and we regard these isomorphisms as identities. Thus, elements of G can be uniquely written as ab, $a, b \in \mathbb{C}$, and $ba = e^b(a)$.

Now let V be a G-module. There are linear endomorphisms X, Y of V with $a(v) = e^{aX}(v)$ and $b(v) = e^{bY}(v)$ for $v \in V$, where a, b have the above significance.

(The Lie algebra of G has basis x, y with brackets $[y, x] = x$. Since G is simply-connected, G-modules and Lie (G) modules are the same. The arguments in the next paragraph show this explicitly.)

Since $ba = e^b(a)$, $e^{bY} e^{aX} e^{-bY} = e^{e^b aX}$. This implies that $e^{bY} aX e^{-bY} = e^b aX$, or $e^{bY} X = e^b X e^{bY}$. Thus, $X + bYX + b^2 W = X + bX + bXY + b^2 U$; subtracting X from both sides, dividing by b, and then setting $b = 0$ shows that $YX = XY + X$, or that $[Y, X] = X$. Conversely, given a vector space V and linear transformations X, Y of V with $[Y, X] = X$, it is easy to prove by induction on k and l that $Y^k X^l = X^l (Y + 1)^k$. Then, $e^{bY} X^l \cdot e^{-bY} = (\sum (k!)^{-1} b^k Y^k X^l) e^{-bY} = X^l (\sum (k!)^{-1} b^k (Y + 1)^k) e^{-bY} = X^l e^{b(Y+1)} e^{-bY} = e^b X^l$ and, hence, $e^{bY} e^{aX} e^{-bY} = e^{e^b aX}$. Then V becomes a G-module with $ab(v) = e^{aX}(e^{bY}(v))$. Thus, $\mathrm{Mod}(G)$ consists of triples (V, X, Y) where X, Y are linear endomorphisms of V with $[Y, X] = X$. A homomorphism $f: (V, X, Y) \to (V', X', Y')$ in $\mathrm{Mod}(G)$ is a linear transformation with $X'f = fX$ and $Y'f = fY$.

Let (V, X, Y) be in $\mathrm{Mod}(G)$. The linear span $\langle X, Y \rangle$ of X and Y in $\mathrm{End}(V)$ is a solvable Lie algebra and, hence, X and Y can be simultaneously put in upper triangular form as matrices, so $X = [Y, X]$ is strictly upper triangular, hence nilpotent as an endomorphism. Now let $Y = D + N$ be the Jordan decomposition of Y: D is a semisimple endomorphism, N is nilpotent, and $DN = ND$. In $\mathrm{gl}(V)$, $\mathrm{ad}(Y) = \mathrm{ad}(D) + \mathrm{ad}(N)$ is the Jordan decomposition of $\mathrm{ad}(Y)$. Since $\mathrm{ad}(Y)X \in \langle X \rangle$ (the linear span of X), $\mathrm{ad}(N)\langle X \rangle \subseteq$

$\langle X \rangle$. Thus, X is an eigenvector of $ad(N)$ and $ad(N)$ is nilpotent, so $ad(N)X = 0$ or $[N, X] = 0$. Thus, N and X are commuting nilpotents and, since $[Y, X] = X$, $[D, X] = X$ also. Of course, D and N still commute. Conversely, given X, N, D in $End(V)$ with X, N commuting nilpotents and D semisimple, D commuting with N and $[D, X] = X$, we can let $Y = D + N$ and (V, X, Y) is in $Mod(G)$. The uniqueness of Jordan decomposition allows the recovery of D and N from Y. Thus, we can regard objects of $Mod(G)$ as quadruples (V, X, N, D) where X, N, D are as above.

We want to consider some special objects in $Mod(G)$: for n, m nonnegative integers and $d \in \mathbb{C}$, let $V_{n,m}(d)$ be the vector space with basis $z_{i,j}$, $0 \le i \le n - 1$, $0 \le j \le m - 1$, and endomorphisms X, N, D defined as follows: $X(z_{i,j}) = z_{i+1,j}$, $N(z_{i,j}) = z_{i,j+1}$ and $D(z_{i,j}) = (d + i)z_{i,j}$. Then X and N are nilpotent, X and N commute, and N and D commute. Also, $(DX - XD)(z_{i,j}) = z_{i+1,j} = X(z_{i,j})$ so $[D, X] = X$ and $V_{n,m}(d)$ is a G-module.

Now let (V, X, N, D) be any G-module, and let $v \in V$ be an eigenvector of D with eigenvalue d. Suppose $X^n(v) = 0$ and $N^m(v) = 0$. Define $f \colon V_{n,m}(d) \to V$ to be the linear transformation with $f(z_{i,j}) = X^i N^j(v)$. Then $f(X(z_{i,j})) = X^{i+1} N^j(v) = Xf(z_{i,j})$, $f(N(z_{i,j})) = X^i N^{j+1}(v) = NX^i N^j(v) = Nf(z_{i,j})$, and $f(D(z_{i,j})) = (d+i)X^i N^j(v)$. Now $D(X^i N^j)(v) = X^i(D+i)N^j(v) = X^i N^j(D+i)(v) = (d+i)X^i N^j(v)$ also, so $f(D(z_{i,j})) = Df(z_{i,j})$. (We have used the formula $DX^i = X^i(D+i)$ which follows by induction from the formula $DX = XD + X = X(D+1)$ which restates the equation $[D, X] = X$.) Thus, f is a morphism in $Mod(G)$, and it is clear that V is a homomorphic image of a direct sum of the $V_{n,m}(d)$ for various n, m and d.

To determine tensor automorphisms of $Mod(G)$, we need only look at the $V_{n,m}(d)$. We examine first the tensor product structure of $V_{n,m}(d)$. Let $V_n(d) = V_{n,1}(d)$ and let $W_m = V_{1,m}(0)$. Let $w_{i,j} = z_{i,0} \otimes z_{0,j}$ in $V_n(d) \otimes W_m$. Then $X(w_{i,j}) = Xz_{i,0} \otimes z_{0,j} + z_{i,0} \otimes Xz_{0,j} = z_{i+1,0} \otimes z_{0,j} = w_{i+1,j}$. Similar calculations show that $N(w_{i,j}) = w_{i,j+1}$ and $D(w_{i,j}) = (d+i)w_{i,j}$. Then $V_n(d) \otimes W_m \to V_{n,m}(d)$ by $w_{i,j} \mapsto z_{i,j}$ is an isomorphism in $Mod(G)$. So we need only consider tensor automorphisms on $V_n(d)$ and W_m separately.

The latter can be handled by the methods of Example C: on W_m, X and D are both zero, and the embedding $W_n \to W_2^{\otimes n}$ used in Example C is a morphism in $\mathrm{Mod}(G)$. Thus, the tensor automorphisms on all W_n are determined by their values on W_2.

When we work with $V_n(d)$, $N = 0$ and we only need consider X and D. We let $V_n = V_n(0)$ and let $v_i = z_{i,0}$ so $X(v_i) = v_{i+1}$ and $D(v_i) = iv_i$, and let $V(d) = V_1(d)$ and $y_0 = z_{0,0}$ so $D(y_0) = dy_0$. Then $V_n \otimes V(d)$ has basis $v_i \otimes y_0$, $i = 0, \ldots, n-1$, with $X(v_i \otimes y_0) = v_{i+1} \otimes y_0$ and $D(v_i \otimes y_0) = (d+i)(v_i \otimes y_0)$, so $V_n \otimes V(d) \to V_n(d)$ by $v_i \otimes y_0 \mapsto z_{i,0} \otimes z_{0,0}$ is an isomorphism. We consider $V_n \otimes V_2$. By induction on k, it is easy to check that $X^k(v_0 \otimes v_0) = v_k \otimes v_0 + k(v_{k-1} \otimes v_1)$, and then that $DX^k(v_0 \otimes v_0) = kX^k(v_0 \otimes v_0)$. The vectors $w_i = X^i(v_0 \otimes v_0)$, $i = 0, \ldots, n$, are linearly independent, satisfy $X(w_i) = w_{i+1}$ and $D(w_i) = iw_i$. Thus, there is a G-module injection $V_{n+1} \to V_n \otimes V_2$ which sends v_i to w_i.

Assembling all the above data, we find that a tensor automorphism is determined by its values on W_2, V_2, and the $V(d)$ for $d \in \mathbb{C}$. As in Example A, we find that $V(c) \otimes V(d) = V(c+d)$. Also, $W_1 = V_1 = V(0)$. W_2 has a basis w_0, w_1 with $N(w_0) = w_1$, $N(w_1) = 0$, and $N = D = 0$; V_2 has basis v_0, v_1 with $X(w_0) = v_1$, $X(y_1) = 0$, $D(v_0) = 0$, $D(v_1) = v_1$ and $N = 0$. To study V_2, we have the exact sequence $V(1) \to V_2 \to V_1 = V(0)$ where the first map sends y_0 to v_1 and the second sends v_0 to y_0 and v_1 to 0. To study W_2, we have the exact sequence $V(0) = W_1 \to W_2 \to W_1 = V(0)$, where the first map sends y_0 to w_1 and the second sends w_0 to y_0 and w_1 to 0.

Now let α be a tensor automorphism of $\mathrm{Mod}(G)$. Let α_v be α on V_2, let α_w be α on W_2 and let α_d be α on $V(d)$. Since $V(d)$ is one-dimensional, $\alpha_d \in \mathbb{C}^*$, and since $V(c) \otimes V(d) = V(c+d)$, $\alpha_c\alpha_d = \alpha_{c+d}$, so α determines an element of $\mathrm{Hom}(G, \mathbb{C}^*)$. Then $\alpha_0 = 1$, so the sequence $V(0) \to W_2 \to V(0)$ shows that $\alpha_w(w_0) = w_0 + bw_1$ and $\alpha_w(w_1) = w_1$ for some $b \in \mathbb{C}$. The sequence $V(1) \to V_2 \to V(0)$ shows that $\alpha_v(v_0) = v_0 + av_1$ for some $a \in \mathbb{C}$, and that $\alpha_v(v_1) = \alpha_1 v_1$.

Now let $g = (a, b)$, which we write ab by our previous conventions. We want to determine $\sigma(g)$ on W_2, V_2 and $V(d)$. On W_2, $X = 0$ and $D = 0$, so $Y = D + N = N$, and $N^2 = 0$, so $e^{aX} = I$ and

$e^{bY} = I + bN$. Thus, $\sigma(g)(w_0) = e^{aX} e^{bY}(w_0) = w_0 + bw_1$ and $\sigma(g)(w_1) = w_1$, so $\sigma(g) = \alpha$ on W_2. On V_2, $N = 0$, so $Y = D$ and $X^2 = 0$. Thus, $e^{bY}(v_0) = v_0$ and $e^{bY}(v_1) = e^b v_1$, while $e^{aX} = I + aX$, so $e^{aX}(v_0) = v_0 + av_1$ and $e^{aX}(v_1) = v_1$. Thus, on V_2, $\sigma(g)(v_0) = e^{aX} e^{bY}(v_0) = v_0 + av_1$ and $\sigma(g)(v_1) = e^b v_1$. On $V(d)$, $X = N = 0$, so $Y = D$ and $e^{aX} = I$, $e^{bY} = e^{bd}I$. Thus, $\alpha = \sigma(g)$ if and only if $\alpha_d = e^{bd}$ for all $d \in \mathbb{C}$.

We note that the calculations so far show that $\sigma \colon G \to \mathrm{Aut}_{\otimes}(\mathrm{Mod}(G))$ is injective. We define $\tau \colon \mathrm{Aut}_{\otimes}(\mathrm{Mod}(G)) \to \mathrm{Hom}(\mathbb{C}, \mathbb{C}^*)$ by $\tau(\alpha)(d) = \alpha_d\, e^{-cd}$, where $cw_1 = \alpha_w(w_0) - w_0$. (The above analysis of w_2 shows that $\alpha \mapsto c$ is a homomorphism $\mathrm{Aut}_{\otimes}(\mathrm{Mod}(G))$ to $\mathbb{C}$.) If $\tau(\alpha) = 1$, then $\alpha_d = e^{cd}$ for all d, and the above calculations show that $\alpha = \sigma(g)$ for suitable g. Also, if $g = ab$, $c = b$ and $\sigma(g)_d = e^{bd}$ so $\tau\sigma(g) = 1$. Thus, we have an exact sequence $1 \to G \to \mathrm{Aut}_{\otimes}(\mathrm{Mod}(G)) \to \mathrm{Hom}(\mathbb{C}, \mathbb{C}^*)$. (This is not the whole story. The general theory to be developed in later chapters will show that τ is surjective and the sequence splits, so $\mathrm{Aut}_{\otimes}(\mathrm{Mod}(G))$ is a semidirect, but not direct, product of the normal subgroup $\sigma(G)$ and $\mathrm{Hom}(\mathbb{C}, \mathbb{C}^*)$. It is worthwhile to consider why we cannot construct a splitting as in Example A: if f is in $\mathrm{Hom}(\mathbb{C}, \mathbb{C}^*)$, why not define a tensor automorphism β by making β the identity on all W_n, V_m and $\beta_d = f(d)$ on $V(d)$? The reason is that the $V(d)$s appear in the V_ns; consider the sequence we used above $V(1) \to V_2 \to V(0)$.)

Example E   Let $\alpha, \beta \in \mathbb{C}$ and let $G = G_{\alpha,\beta}$ be the semidirect product of $\mathbb{C}^{(2)}$ with $\mathbb{C}$, where the latter acts on the former via $s \colon \mathbb{C} \to \mathrm{GL}_2\mathbb{C}$ where $s(c) = \mathrm{diag}(e^{\alpha c}, e^{\beta c})$; so $G = \{(a, b, c) \mid a, b, c \in \mathbb{C}\}$ and $(a, b, c)(a', b', c') = (a + e^{\alpha c}a', b + e^{\beta c}b', c + c')$.

Our aim in this example is not to compute $\mathrm{Aut}_{\otimes}(\mathrm{Mod}(G))$ (which we could determine later from our general theory), but to point out the independence, in a sense, of $\mathrm{Mod}(G)$ from the choice of α and β. We will use this example to see that nonisomorphic groups can have equivalent categories of modules.

We begin as in Example D. The element (a, b, c) will be written as abc, where $a \in \mathbb{C} = \mathbb{C} \times \{0, 0\}$, $b \in \mathbb{C} = \{0\} \times \mathbb{C} \times \{0\}$, and $c \in \mathbb{C} = \{0, 0\} \times \mathbb{C}$. Then $ca = e^{\alpha c}(a)$ and $cb = e^{\beta c}(b)$, and these relations show how to compute all products in G.

Now let V be a G-module. There are linear endomorphisms X, Y, Z of V with $a(v) = e^{aX}(v)$, $b(v) = e^{bY}(v)$, and $c(v) = e^{cZ}(v)$, where $v \in V$ and a, b, c in G have the above significance.

The reader can easily verify, by the techniques of Example D, that these linear endomorphisms have the following commutation relations: $[Z, X] = \alpha X$, $[Z, Y] = \beta Y$ and $[X, Y] = 0$. Conversely, given a vector space V and linear endomorphisms satisfying these relations, V becomes a G-module under $abc(v) = e^{aX} e^{bY} e^{cZ}(v)$. Thus, we can think of G-modules as quadruples (V, X, Y, Z). Again as in Example D, we take the Jordan decomposition $Z = D + N$ where D and N commute, D is semisimple and N is nilpotent. Thus, G-modules are quintuples (V, X, Y, D, N) where V is a vector space, X, Y, D, N are endomorphisms of V with D semisimple, N nilpotent and the following commutation relations hold: $[D, X] = \alpha X$, $[D, Y] = \beta Y$, $[D, N] = [N, X] = [N, Y] = [X, Y] = 0$.

If (V, X, Y, D, N) and (W, X', Y', D', N') are objects of $\mathrm{Mod}(G)$, then a morphism f from the first to the second is a linear transformation $f: V \to W$ with $T'f = fT$ for $T = X, Y, D, N$ and their tensor product is $(V \otimes W, X'', Y'', D'', N'')$ where $T''(v \otimes w) = Tx \otimes w + x \otimes T'w$ for $T = X, Y, D, N$.

We now want to see how the category $\mathrm{Mod}(G)$ changes under certain changes in α and β. Now α and β are only involved in the commutation relations with D, X and Y, but a change in D also affects the commutation relations with N. We want to express all these relations in terms of eigenspaces of D. Thus, let (V, X, Y, D, N) be in $\mathrm{Mod}(G)$. For $d \in \mathbb{C}$, let $V_d = \{v \in V \mid D(v) = dv\}$. Since $\alpha X = DX - XD$ or $DX = X(\alpha + D)$, if $v \in V_d$, $DXv = (\alpha + d)Xv$, so $X(V_d) \subseteq V_{d+\alpha}$. Conversely, if $X(V_d) \subseteq V_{d+\alpha}$ for all d, then $DX = X(\alpha + D)$ or $[D, X] = \alpha X$. Similarly, the relation $[D, Y] = \beta Y$ is equivalent to $Y(V_d) \subseteq V_{d+\beta}$ for all d, and the relation $[D, N] = 0$ is equivalent to $N(V_d) \subseteq V_d$ for all d.

Now, let $h: \mathbb{C} \to \mathbb{C}$ be an additive group automorphism of $\mathbb{C}$. We emphasize that h is *not* necessarily continuous. Let $\alpha' = h(\alpha)$, $\beta' = h(\beta)$ and $G' = G_{\alpha', \beta'}$. Let $V = (V, X, Y, D, N)$ be in $\mathrm{Mod}(G)$. We are going to modify V to produce a G'-module $h_* V = (V, X, Y, D', N)$ where X, Y, N are as before and D' is defined as follows: if $v \in V_d$, $D'(v) = h(d)v$. We have that $[N, X] = [N, Y] = [X, Y] = 0$, D' is semisimple and N is nilpotent. Let $V'_d = \{v \in V \mid D'(v) = dv\}$, so $V'_d = V_{h^{-1}(d)}$. Then since $X(V_d) \subseteq V_{d+\alpha}$, $X(V'_d) \subseteq V_{h^{-1}(d)+\alpha} = V_{h^{-1}(d+\alpha')} = V'_{d+\alpha'}$. Similarly, $Y(V'_d) \subseteq V'_{d+\beta'}$ and $N(V'_d) \subseteq V'_d$. As we just noted, these inclusions imply the commutator relations $[D', X] = \alpha'X$, $[D', Y] = \beta'Y$ and $[D', N] = 0$. Thus, $h_* V$ is a G'-module.

If $f: V \to W$ is a morphism in Mod(G), define $h_* f: h_* V \to h_* W$ as $h_* f = f$. Since $f(V_d) \subseteq W_d$, $h_* f$ turns out to be a G'-morphism as the reader can quickly verify. Also, $V_i = (V_i, X_i, Y_i, D_i, N_i)$ for $i = 1, 2$ are in Mod(G), then $h_*(V_1 \otimes V_2) = h_* V_1 \otimes h_* V_2$ in Mod(G'): for let $V_1 \otimes V_2 = (V_1 \otimes V_2, X, Y, D, N)$. If $d, e \in \mathbb{C}$ and $v \in (V_1)_d$ and $w \in (V_2)_e$, then $D(v \otimes w) = D_1 v \otimes w + v \otimes D_2 w = (d + e)(v \otimes w)$, so $(V_1)_d \otimes (V_2)_e \subseteq (V_1 \otimes V_2)_{d+e}$. So if $v \in (V_1)'_d$ and $w \in (V_2)'_e$, then $v \otimes w \in (V_1 \otimes V_2)_{h^{-1}(d+e)}$, so $D'(v \otimes w) = h^{-1}(d + e)(v \otimes w) = h^{-1}(d)v \otimes w + v \otimes h^{-1}(e)w = (D'_1 v) \otimes w + v \otimes (D'_2 w)$. This shows that D' arises in the proper fashion from D'_1 and D'_2, and the transformations X, Y, N are unchanged.

In summary, the automorphism h induces a functor $h_*: \text{Mod}(G) \to \text{Mod}(G')$ which preserves tensor products. Of course, h^{-1} also induces a functor $(h^{-1})_*: \text{Mod}(G') \to \text{Mod}(G)$ inverse to h_*, so that h_* is an equivalence of tensored categories. Among other things, this implies that $\text{Aut}_\otimes(\text{Mod}(G))$ is isomorphic to $\text{Aut}_\otimes(\text{Mod}(G'))$.

We now want to show how to choose α, β, and h so that G and G' are not isomorphic as complex Lie groups. This is easier to do it we pass to Lie algebras. Lie(G) has basis x, y, z with multiplication $zx = \alpha x$, $zy = \beta y$ and $xy = 0$. If $w = vx + sy + tz$ is any element of Lie(G), then $\text{ad}(w)(x) = t\alpha x$, $\text{ad}(w)(y) = t\beta y$, and $\text{ad}(w)(z) = -v\alpha x - s\beta y$. It follows that the eigenvalues of w are $\{t\alpha, t\beta\}$. Now assume both α and β are nonzero. Then the ratio of the eigenvalues of any element of Lie(G) with nonzero eigenvalues is $\alpha\beta^{-1}$ or $\alpha^{-1}\beta$. If, for example, we choose $\beta = 1$, then the pair $\{\alpha, \alpha^{-1}\}$ is determined by Lie(G). If we take 1, α, and α' in $\mathbb{C}$ linearly independent over $\mathbb{Q}$ with $\alpha' \neq \alpha^{-1}$, then we can choose an automorphism h of $\mathbb{C}$ (as $\mathbb{Q}$-vector space) with $h(1) = 1$ and $h(\alpha) = \alpha'$. Then $G = G_{\alpha,1}$ and $G' = G_{\alpha',1}$ are not isomorphic (since

$\{\alpha, \alpha^{-1}\} \neq \{\alpha', (\alpha')^{-1}\}$, but $h_*: \mathrm{Mod}(G) \to \mathrm{Mod}(G')$ is an equivalence of tensored categories.

Summary remarks on the examples of Chapter 1

The methods used to describe the categories of modules, and their tensor automorphisms, in the examples are *ad hoc*. (We will see later that there are systematic methods we can use in general, however, which yield equivalent descriptions.) Nonetheless, we can at least observe some common phenomena at this point. In the examples where we succeeded in determining $\mathrm{Aut}_{\otimes}(\mathrm{Mod}(G))$, the description of $\mathrm{Mod}(G)$ took the following form: we produced all modules as submodules or quotients of tensor products of some infinite families of one-dimensional modules ($V(a)$ in Example B, $V(d)$ in Example D) and a finite set of modules ($\mathbb{C}[1]$ in Example A, V_2 in Example B, W_2 and $W[1]$ in Example C, W_2 and V_2 in Example D). Moreover, it was in the examples (B and D) where σ was not an isomorphism that these infinite families appeared.

It will turn out that this is symptomatic: the groups G for which σ is not onto are exactly those for which the category of one-dimensional modules is not finitely-generated as a tensored category, i.e., for which not every one-dimensional module is a tensor product of one-dimensional modules in some fixed finite set.

We also observe that in all the examples $\mathrm{Aut}_{\otimes}(\mathrm{Mod}(G))$ was an extension of a subgroup of $\mathrm{Hom}(\mathbb{C}, \mathbb{C}^*)$ by $\sigma(G)$. It will turn out in general that $\mathrm{Aut}_{\otimes}(\mathrm{Mod}(G))$ is the semidirect product of the normal subgroup $\sigma(G)$ and a group $\mathrm{Hom}(\mathbb{C}^{(n)}, \mathbb{C}^*)$, where n is the number of linearly independent analytic homomorphisms from G to $\mathbb{C}$. Thus, we will have $\sigma(g) = \mathrm{Aut}_{\otimes}(\mathrm{Mod}(G))$ if and only if $n = 0$, and we note that this is the case for Examples A and C. Moreover, the groups of A and C are algebraic.

As we pointed out at the end of Example D the semidirect product $\mathrm{Aut}_{\otimes}(\mathrm{Mod}(G)) = \sigma(G)\mathrm{Hom}(\mathbb{C}, \mathbb{C}^*)$ is not direct, which it was in Example B. Now the group in Example B is algebraic while that of Example D is not, and it turns out that in general the semidirect product is direct if and only if G is algebraic.

The reader may wonder why we want to use three-dimensional groups to give an example of nonisomorphic groups with equivalent module categories in Example E. The reason, of course, is that three is minimal, although, again, it is not possible to demonstrate this in an elementary fashion.

Finally, all the examples were of solvable groups. We will see later that for reductive groups σ is always an isomorphism (this follows from the fact, already mentioned, that $\sigma(G) =$ $\mathrm{Aut}_\otimes(\mathrm{Mod}(G))$ if there are no analytic homomorphisms from G to $\mathbb{C}$).

2

Representative functions

This chapter introduces the Hopf algebra of representative analytic functions on an analytic group. There is a categorical equivalence between modules for the group and finite-dimensional comodules for the algebra. From this equivalence comes an identification of the group of tensor automorphisms of the category of modules for the analytic group with the group of $\mathbb{C}$-valued $\mathbb{C}$-algebra homomorphisms of the Hopf algebra. This latter group has a representation theory equivalent to the original analytic group. The material in this chapter can be regarded as an 'algebrization' of the representation theory of an analytic group.

We will start with the notion of a representative function, which could be defined as a matrix coefficient function of a representation. We prefer to give an intrinsic definition (which is essentially equivalent, see (2.7) below). We will need the following basic lemma.

Lemma 2.1 Let X be a set and let V be a finite-dimensional vector space of $\mathbb{C}$-valued functions on X. Then there is a basis $f_1, \ldots, f_d$ of V and elements $x_1, \ldots, x_d$ of X such that $f_i(x_j) = \delta_{ij}$ for $1 \le i, j \le d$.

Proof. For each $x \in X$, let $e_x : V \to \mathbb{C}$ be given by $e_x(f) = f(x)$. Then the e_x are $\mathbb{C}$-linear (so $e_x \in V^*$) and $\{e_x \mid x \in X\}^\perp = \{f \in V \mid e_x(f) = 0$ for all $x \in X\} = \{0\}$, so $\{e_x \mid x \in X\}$ contains a basis $e_{x_1}, \ldots, e_{x_d}$ of V^*. Let $f_1, \ldots, f_d$ be the basis of V dual to $e_{x_1}, \ldots, e_{x_d}$. Then, $f_i(x_j) = e_{x_j}(f_i) = \delta_{ij}$.

Definition 2.2 Let G be a group, $f : G \to \mathbb{C}$ a function, and $x \in G$. Then $f \cdot x : G \to \mathbb{C}$ (respectively, $x \cdot f : G \to \mathbb{C}$) is the function given by $(f \cdot x)(g) = f(xg)$ (respectively, $(x \cdot f)(g) = f(gx)$). Then the fol-

32

lowing relations hold (for $x, y \in G$, $f, g : G \to \mathbb{C}$ and $a \in \mathbb{C}$):
$(f \cdot x) \cdot y = f \cdot (xy)$, $x \cdot (y \cdot f) = (xy) \cdot f$ and $(x \cdot f) \cdot y = x \cdot (f \cdot y)$;
$x \cdot (af + g) = a(x \cdot f) + x \cdot g$ and $(af + g) \cdot y = a(f \cdot y) + g \cdot y$;
$x \cdot (fg) = (x \cdot f)(x \cdot g)$ and $(fg) \cdot x = (f \cdot x)(g \cdot x)$.

If the group G in (2.2) is analytic, and f is an analytic function on G, then $x \cdot f$ and $f \cdot x$ are also analytic functions on G.

Definition 2.3 Let G be a group and let $f : G \to \mathbb{C}$ be a function. Then:

  (a) $[f]_r = \langle f \cdot x \, | \, x \in G \rangle$;
  (b) $_l[f] = \langle x \cdot f \, | \, x \in G \rangle$;
  (c) $_l[f]_r = \langle x \cdot f \cdot y \, | \, x \in G \rangle$.

Lemma 2.4 Let G be a group and let $f : G \to \mathbb{C}$ be a function. Then the following are equivalent:

  (a) $[f]_r$ is finite-dimensional,
  (b) $_l[f]$ is finite-dimensional;
  (c) $_l[f]_r$ is finite-dimensional;
  (d) There are functions $h_i, k_i : G \to \mathbb{C}$, $1 \le i \le n$, such that $f(xy) = \sum h_i(x)k_i(y)$ for all $x, y \in G$. Moreover, we can take $h_i \in {}_l[f]$ and $k_i \in [f]_r$.

Proof. Assume (a) holds. We want to choose a basis of $[f]_r$. By (2.1), we can find a basis $k_1, \ldots, k_d$ of $[f]_r$ and elements $y_1, \ldots, y_d$ of G such that $k_i(y_j) = \delta_{ij}$. If $x \in G$, then $f \cdot x \in [f]_r$, so we can write $f \cdot x = \sum h_i(x)k_i$ with $h_i(x) \in \mathbb{C}$. Then $(f \cdot x)(y_j) = \sum h_i(x)k_i(y_j) = h_j(x)$, or $h_j(x) = (f \cdot x)(y_j) = f(xy_j) = (y_j \cdot f)(x)$. Hence, $h_j = y_j \cdot f$, and we have $f(xy) = \sum h_i(x)k_i(y)$ with $h_i \in {}_l[f]$ and $k_i \in [f]_r$. Thus (a) implies (d).

  An exactly parallel argument shows that (b) implies (d).

  Now assume that (d) holds. Then $f \cdot x = \sum h_i(x)k_i$ so $[f]_r \subseteq \langle k_1, \ldots, k_n \rangle$ and $y \cdot f = \sum k_i(y)h_i$ so $_l[f] \subseteq \langle h_1, \ldots, h_n \rangle$, and thus $[f]_r$ and $_l[f]$ are finite-dimensional. Thus (d) implies (a) and (b).

  If (c) holds, then so do (a) and (b) since $[f]_r \subseteq {}_l[f]_r$ and $_l[f] \subseteq {}_l[f]_r$. To complete the proof of the lemma, we will show that (d) implies (c). Thus, assume (d) holds. This implies that (a) holds, so there are $x_1, \ldots, x_m$ in G such that $[f]_r = \langle f \cdot x_1, \ldots, f \cdot x_m \rangle$. Then $_l[f]_r \subseteq \sum {}_l[f \cdot x_i]$, so it suffices to prove that $_l[f \cdot x_i]$ is finite-dimensional. But $[f \cdot x_i]_r = [f]_r$ is finite-dimensional, so (d) holds for $f \cdot x_i$, and, hence, (b) holds for $f \cdot x_i$, so $_l[f \cdot x_i]$ is finite-dimensional also.

We can now define representative function.

Definition 2.5 Let G be an analytic group. An analytic function f on G is a *representative function* if f satisfies the equivalent conditions of (2.4).

Let us note that a representative function is analytic by definition. Moreover, it follows from (2.4) that if f is representative, so are $x \cdot f$ and $f \cdot x$.

Now let f be a representative function on the analytic group G. Then $[f]_r$ is a finite-dimensional vector space, and we can give it a G-action by the formula $x(v) = v \cdot x^{-1}$ for $v \in [f]_r$ and $x \in G$. We want to see that this action makes $[f]_r$ an (analytic) G-module. Suppose we have chosen a basis $k_1, \ldots, k_d$ of $[f]_r$ such that there are elements $y_1, \ldots, y_d$ of G with $k_i(y_j) = \delta_{ij}$, as we can do by (2.1). Then the proof that (a) implies (d) of (2.4) showed that $f \cdot x = \sum h_i(x)k_i$, where $h_i = y_i \cdot f$. The equation $g \cdot x = \sum (y_i \cdot g)(x)k_i$ then will hold for any element g of $[f]_r$, since $[g]_r \subseteq [f]_r = \langle k_1, \ldots, k_d \rangle$, so, in particular we have, for each i, $x(k_i) = k_i \cdot x^{-1} = \sum (y_i \cdot k_i)(x^{-1})k_j$. Thus, the matrix representing x on $[f]_r$ in the basis $k_1, \ldots, k_d$ has (i, j)-entry $(y_j \cdot k_i)(x^{-1})$, and this is an analytic function of x.

In a similar fashion, we can make $_l[f]$ an analytic G-module with action $x(v) = x \cdot v$ for x in G and $v \in {}_l[f]$. For future reference we record these facts.

Proposition 2.6 Let G be an analytic group and f a representative function on G. Then $[f]_r$ (respectively, $_l[f]$) becomes a G-module under the action $x(v) = v \cdot x^{-1}$ (respectively, $x(v) = x \cdot v$) for x in G and v in $[f]_r$ (respectively, $_l[f]$).

We have just seen how representative functions lead to modules. Conversely, we shall also see how modules yield representative functions.

Proposition 2.7 Let G be an analytic group and V an object of Mod(G). If $v \in V$ and $\phi \in V^*$, then the function $f(x) = \phi(x(v))$ is a representative function on G.

Proof. Obviously, f is analytic. Let $v_1, \ldots, v_n$ be a basis of V, and define $f_i : G \to \mathbb{C}$ by $f_i(x) = \phi(x(v_i))$. We also define $\alpha_i : G \to \mathbb{C}$ by $\sum \alpha_i(y)v_i = y(v)$. Now $(f \cdot x)(y) = f(xy) = \phi(xy(v)) = \sum \alpha_i(y)\phi(x(v_i)) = \sum \alpha_i(y)f_i(x)$. Thus, $f \cdot x \in \langle \alpha_1, \ldots, \alpha_n \rangle$ for all x, so $[f]_r \subseteq \langle \alpha_1, \ldots, \alpha_n \rangle$ and f is representative.

Notation 2.8 In the situation of (2.7), we will sometimes denote the function f as $f_{\phi,v}$ to emphasize the dependence on ϕ and v.

Remark. Actually, every representative function on G has the form of (2.7): if f is a representative function on G and we let $v \in {}_l[f]$ be f and $\phi \in ({}_l[f])^*$ be evaluation at e, then $(f_{\phi,v})(x) = \phi(x \cdot f) = (x \cdot f)(e) = f(x)$.

Proposition 2.7 helps explain the language 'representative function': if $\rho : G \to \mathrm{GL}(V)$ is an (analytic) representation of G, and we choose a basis $v_1, \ldots, v_n$ of G with dual basis $\phi_1, \ldots, \phi_n$, then the (i, j)-entry of the matrix of $\rho(x)$ in the basis $v_1, \ldots, v_n$ is $\phi_j(\rho(x)v_i) = f_{\phi_j, v_i}(x)$. So the coefficient functions of the matrix representation obtained from ρ using the chosen basis are representative functions.

There are some simple but useful relations among the functions $f_{\phi,v}$ of (2.7) which we now record. The easy verifications are left to the reader.

Lemma 2.9 Let G be an analytic group and V a G-module. Let ϕ, ψ be in V^*, v, w in V, a in $\mathbb{C}$ and x in G. Then:

(a) $f_{\phi+\psi,v} = f_{\phi,v} + f_{\psi,v}$;

(b) $f_{\phi,v+w} = f_{\phi,v} + f_{\phi,w}$;

(c) $f_{a\phi,v} = af_{\phi,v} = f_{\phi,av}$;

(d) $f_{x\phi,v} = f_{\phi,v} \cdot x^{-1}$;

(e) $f_{\phi,xv} = x \cdot f_{\phi,v}$.

We will use the special representative functions $f_{\phi,v}$ to show below, (2.14), that every 'abstract' G-module is isomorphic to a G-module of representative functions. So that the discussion can be complete, we need to consider some properties of the totality of representative functions.

Notation 2.10 Let G be an analytic group. The set of all representative functions on G is denoted $R(G)$.

Lemma 2.11 Let G be an analytic group. Then $R(G)$ is a commutative $\mathbb{C}$-algebra with identity and no zero-divisors, under pointwise operations.

Proof. To see that $R(G)$ is a ring, we use $(2.4)(a)$ and the following easily verified inclusions: if $f, g \in R(G)$, then $[f+g]_r \subseteq [f]_r + [g]_r$ and $[fg]_r \subseteq [f]_r[g]_r$. The constant functions are in $R(G)$ and the constant 1 is an identity for $R(G)$. If $a \in \mathbb{C}$ and $f \in R(G)$, $[af]_r \subseteq [f]_r$ (with equality if $a \neq 0$), so $R(G)$ is a $\mathbb{C}$-algebra. The fact that $R(G)$ has no zero-divisors follows from the fact that G is connected and the elements of $R(G)$ are analytic functions on G.

There are some G-modules inside $R(G)$, for example, those of (2.6). We introduce some language to refer to such modules conveniently.

Definition 2.12 Let G be an analytic group. A subspace V of $R(G)$ is called *left stable* (respectively, *right stable*) if for all f in V and x in G, $x \cdot f$ is in V (respectively, $f \cdot x$ is in V).

Lemma 2.13 Let G be an analytic group and V a finite-dimensional left stable (respectively, right stable) subspace of $R(G)$. Then V is a G-module under the action $x(v) = x \cdot v$ (respectively, $x(v) = v \cdot x^{-1}$).

Proof. The verification that the stated formulae give actions follows from the equations in (2.2). To verify that the actions give G-modules we assume that V is left stable with basis $k_1, \ldots, k_n$. For each i, $_l[k_i] \subseteq V$ is an inclusion of vector spaces with G-action, so $\prod_l[k_i] \to V$ is a surjection of vector spaces with G-action. The domain is a G-module by (2.6) and, hence, V is a quotient of G-modules, hence, a G-module. The argument is similar for right stable subspaces.

Now suppose V is any G-module and ϕ is in V^*. We can define a function $T_\phi : V \to R(G)$ by $T_\phi(v) = f_{\phi,v}$; T_ϕ is linear by (2.9)(b) and (c) and we let W_ϕ denote its image. Now $T_\phi(xv) = x \cdot T_\phi(v)$ by (2.9)(c), so W_ϕ is left stable (and, finite-dimensional) and, hence, a G-module by (2.13). If $T_\phi(v) = 0$, $0 = f_{\phi,v}(e) = \phi(v)$, so v is in the kernel of ϕ. Now suppose $\phi_1, \ldots, \phi_n$ is a basis of V^*. Let $T = (T_{\phi_1}, \ldots, T_{\phi_n})$ so $T : V \to \prod W_{\phi_i}$. Then, $\mathrm{Ker}(T) \subseteq \bigcap \mathrm{Ker}(\phi_i) = 0$, so T is an injection of the G-module V into a G-module of representative functions, which is a direct sum of left stable G-modules in $R(G)$.

Conversely, we could fix v in V and define $S_v : V^* \to R(G)$ by $S_v(\phi) = f_{\phi,v}$. Then (2.9)(a) and (c) show that S_v is linear, and $S_v(x\phi) = S_v(\phi) \cdot x^{-1}$ by (2.9)(d), so the image A_v of S_v is a finite-dimensional right stable subspace of $R(G)$ and, hence, a G-module. Again, $S_v(\phi) = 0$ implies $0 = f_{\phi,v}(e) = \phi(v)$, so $v \in \mathrm{Ker}(\phi)$. If $v_1, \ldots, v_n$ is a basis of V, then $S = (S_{v_1}, \ldots, S_{v_n}) : V^* \to \prod U_{v_i}$ is a G-module injection of V^* into a G-module of representative functions which is a direct sum of right stable subspaces of $R(G)$. If we think $V = (V^*)^*$ and apply this construction with V^* replacing V, we get that V is (isomorphic to) a submodule of a direct sum of right stable subspaces of $R(G)$. We summarize these results in the following theorem.

Theorem 2.14 Let G be an analytic group and V a G-module. Then V is isomorphic to a submodule of a G-module of representative functions which is a direct sum of finite-dimensional left (or right) stable subspaces of $R(G)$.

We can use (2.14) and the algebra structure of $R(G)$ to find 'generators' for the category $\mathrm{Mod}(G)$. If V and W are left (or right) stable finite-dimensional subspaces of $R(G)$, then so are $V + W$ and $VW = \langle vw \mid v \in V, w \in W \rangle$, and the maps $V \oplus W \to V + W$ by $(v, w) \mapsto v + w$ and $V \otimes W \to VW$ by $v \otimes w \mapsto vw$ are G-module epimorphisms. If we apply this observation when V and W are of the form $_l[f]$ (or $[f]_r$) where f is part of a system of algebra generators for $R(G)$ and use (2.14), we conclude the following corollary.

Corollary 2.15 Let G be an analytic group and let $\{f_i \mid i \in I\}$ be a set of algebra generators for $R(G)$. Then every G-module is a submodule of a G-module obtained by tensor products and direct sums from the set of G-modules $\{_l[f_i] \mid i \in I\}$ (and similarly for the set $\{[f_i]_r \mid i \in I\}$).

At this point we interrupt our general discussion to compute $R(G)$ where G is as in Examples A and B of Chapter 1.

Example A' Let $G = \mathrm{GL}_1\mathbb{C}$. We first want to construct a representative function on G. Let $t : G \to \mathbb{C}$ be given by $t(x) = x$. If $y \in G$, $(t \cdot y)(x) = yx = yt(x)$, so $t \cdot y = yt$ so $[t]_r = \langle t \rangle$. Thus, t is representative. Now let f be any representative function on G. We know by the remark following (2.8) that there is a G-module V and $v \in V$, $\phi \in V^*$ with $f = f_{\phi,v}$. As noted in Example A, V has a basis $v_1, \ldots, v_n$ of simultaneous eigenvectors for G, so that $x(v_i) = x^{m_i}v_i$ for some integer m_i. Now let $\phi_1, \ldots, \phi_n$ be the dual basis to $v_1, \ldots, v_n$. Then $f_{i,j} = f_{\phi_i, v_j}$ is the representative function $f_{i,j}(x) = \phi_i(x(v_j)) = \phi_i(x^{m_j}v_j)$, which is x^{m_i} if $i = j$ and zero otherwise. Thus, $f_{i,j} = \delta_{ij}t^{m_i}$. Now if $\phi = \sum a_i\phi_i$ and $v = \sum b_j\, v_j$, then repeated applications of $(2.9)(a)$, (b) and (c) show that $f = \sum a_ib_jf_{i,j}$. Thus, f is a polynomial, with complex coefficients, in t and t^{-1}. We already know that t is representative and a similar calculation shows that $t^{-1} \cdot y = yt^{-1}$ for all y in G, so t^{-1} is representative. Thus, $R(G) = \mathbb{C}[t, t^{-1}]$. We also have that $[t]_r = \langle t \rangle$ and $[t^{-1}]_r = \langle t^{-1} \rangle$. By (2.15), we get all G-modules from $\{[t]_r, [t^{-1}]_r\}$ by submodules of tensor products and direct sums of $[t]_r$ and $[t^{-1}]_r$. In the notation of Example A, if $m \in \mathbb{Z}$, then if $m \geq 0$, $\mathbb{C}[m] = [t]_r^{\otimes m}$ and if $m < 0$, $\mathbb{C}[m] = [t^{-1}]_r^{\otimes -m}$, and every G-module is a direct sum of some $\mathbb{C}[m]$s.

Example B' Let $G = \mathbb{C}$ (additive group). We first list some representative functions on G. Let $t : G \to \mathbb{C}$ be given by $t(x) = x$. If $y \in G$, $(t \cdot y)(x) = t(y + x) = y + x$, so $t \cdot y = y + t$. Thus, $[t]_r \subseteq \langle 1, t \rangle$, so t is in $R(G)$. If $a \in \mathbb{C}$, let $s_a : G \to \mathbb{C}$ be given by $s_a(x) = e^{ax}$. Then, $(s_a \cdot y)(x) = e^{a(y+x)} = e^{ay}s_a(x)$, so $[s_a]_r \subseteq \langle s_a \rangle$ and s_a is in $R(G)$. As in Example A', we know that an arbitrary element f of $R(G)$ is of the form $f = f_{\phi,v}$ where $\phi \in V^*$ and $v \in V$ for some G-module V. From Example A, we know that the G-action on V is given by $x(v) = e^{Ax}(v)$, for some endomorphism A of V. Let $A = D + N$ be the Jordan decomposition of A, where D is semisimple, N is nilpotent, and $DN = ND$. We can choose a basis $v_1, \ldots, v_n$ of V consisting of eigenvectors of D, with $D(v_i) = a_i v_i$, and we let $\phi_1, \ldots, \phi_n$ be the dual basis to $v_1, \ldots, v_n$. We let $f_{i,j} = f_{\phi_i, v_j}$. Then, $x(v_j) = e^{Ax}(v_j) = e^{Nx} e^{Dx}(v_j) = e^{d_j x}(e^{Nx}(v_j))$. Now $N^n = 0$, so $e^{Nx} = I + xN + \cdots + [x^{n-1}/(n-1)!]N^{n-1}$. Thus, $\phi_i(e^{Nx}v_j) = \phi_i(v_j) + x\phi_i(Nv_j) + \cdots + [x^{n-1}/(n-1)!]\phi_i(N^{n-1}v_j)$. Let $\alpha_{i,j,k} = \phi_i[(x^k/k!)N^k v_j]$. Then, $f_{i,j}(x) = \phi_i(x(v_j)) = e^{d_j x}(\sum_{k=0}^{n-1} \alpha_{i,j,k} x^k)$, so $f_{i,j} = s_{d_j}(\sum a_{i,j,k} t^k) = s_{d_j} p_{i,j}(t)$, where $p_{i,j}$ is the polynomial with coefficients $\alpha_{i,j,0}, \ldots, \alpha_{i,j,n-1}$. Finally, if $\phi = \sum a_i \phi_i$ and $v = \sum b_j v_j$, then by (2.9), $f = \sum a_i b_j f_{i,j} = \sum a_i b_j s_{d_j} p_{i,j}(t)$. It follows that $\{t, s_a \mid a \in \mathbb{C}\}$ gives a set of algebra generators for $R(G)$.

We want to explore some of the structure of $R(G)$ that can be deduced from this set of generators. First, $R(G)$ contains the polynomial ring $\mathbb{C}[t]$ and, hence, we can regard $R(G)$ as a $\mathbb{C}[t]$-module. Also, $R(G)$ contains the set $Q = \{s_a \mid a \in \mathbb{C}\}$, and our description of f above shows that Q generates $R(G)$ as a $\mathbb{C}[t]$-module. We are going to show that actually Q freely generates $R(G)$ as a $\mathbb{C}[t]$-module. This requires a certain amount of calculation with complex exponentials.

If p is an analytic function on $\mathbb{C}$ and $a \in \mathbb{C}$, then it is easy to prove by induction that the nth-derivative of $p\, e^{az}$ is given by $(p\, e^{az})^{(n)} = (\sum_{k=0}^{n} \binom{n}{k} a^{n-k} p^{(k)}) e^{az}$. Now suppose $p_1, \ldots, p_n$ are polynomials, $a_1, \ldots, a_n$ are distinct elements of $\mathbb{C}$, and $\sum p_i\, e^{a_i z} = 0$. We want to prove by induction on n that $p_1 = p_i = \cdots = p_n = 0$. This is trivial for the case $n = 1$, so assume by induction it is true for $n - 1$. After relabeling, we can assume $\deg(p_i) \le \deg(p_n)$ for $i = 1, \ldots, n$. Then, $\sum^n p_i\, e^{a_i z} = 0$ implies $\sum^{n-1} p_i\, e^{b_i z} = -p_n$ where $b_i = a_i - a_n$. Let $m = \deg(p_n) + 1$. Then $(-p_n)^{(m)} = 0$, so $0 = \sum^{n-1} (p_i\, e^{b_i z})^{(m)} = \sum^{n-1} (\sum^m \binom{m}{k} b_i^{m-k} p_i^{(k)}) e^{b_i z} = 0$ by our formula above. By induction, $\sum \binom{m}{k} b_i^{m-k} p_i^{(k)} = 0$. If the highest term in p_i is $a_i x^{m_i}$, then this expression gives $b_i^m a_i x^{m_i} + \text{lower degree} = 0$. Since $b_i = a_i - a_n \ne 0$, we have $p_i = 0$.

The calculations of the preceding paragraph show that Q freely generates $R(G)$ as a $\mathbb{C}[t]$-module. For later use, we want to remark on an alternate description of Q: $Q = \exp(X^+(G))$. Thus, Q is an abelian group isomorphic to the additive group $X^+(G)$, which is in turn isomorphic to $\mathbb{C}$.

We also want to use (2.15) to describe $\text{Mod}(G)$: every G-module is a submodule of a G-module obtained by tensor products and direct sums from the G-modules $\{{}_1[t], {}_1[s_a]\,|\,a \in \mathbb{C}\}$. Now ${}_1[s_a] = \langle s_a \rangle$, and if x is in G, $x(s_a) = x \cdot s_a = e^{ax} s_a$. It follows that, in the notation of Example B, that ${}_1[s_a] = V(a)$. Also, ${}_1[t] = \langle 1, t \rangle$ and $x(t) = x \cdot t = x \cdot 1 + t$. Using the notation of Example B, we get a G-isomorphism $f: V_2 \to {}_1[t]$ by $f(ay_0 + by_1) = a \cdot 1 + b \cdot t$. Then all G-modules are obtained by direct sums and tensor products then submodules from the set $\{V_2, V(a)\,|\,a \in \mathbb{C}\}$. When we did this in Example B, we used quotient modules instead of submodules. To help explain how that worked, we will observe that our set of modules is self-dual: it is easy to check that $(V(a))^* = V(-a)$. If ϕ_0 and ϕ_1 are the dual basis to the basis y_0, y_1 of V_2, then since $x(y_0) = y_0$ and $x(y_1) = xy_0 + y$, for x in G, a calculation shows that $x(\phi_0) = \phi_0 - x\phi_1$ and $x(\phi_1) = \phi_1$. Thus we get a G-isomorphism $h: V_2 \to V_2^*$ given by $h(an_0 + bn_1) = a(-\phi_1) + b\phi_0$.

We now return to the study of $R(G)$ for an arbitrary analytic group G. So far, we have considered $R(G)$ with two structures: its algebra structure, and the left and right G-actions. We want to express the G-actions algebraically also, and it turns out that this will be possible when we view $R(G)$ as a Hopf algebra with antipode, which is our next task.

Definition 2.16 If f is any function on the group H, let f' be the function on H defined by $f'(x) = f(x^{-1})$.

If G is a group, $f: G \to \mathbb{C}$ a function and $x \in G$, then $(x \cdot f)' = f' \cdot x^{-1}$ and $(f \cdot x)' = x^{-1} \cdot f'$. It follows that $({}_1[f])' = [f']_r$ and $([f]_r)' = {}_1[f']$.

Lemma 2.17 Let G be an analytic group and let $f \in R(G)$. Then $f' \in R(G)$.

Proof. By (2.4)(*d*), there are functions h_i, $k_i : G \to \mathbb{C}$, $i = 1, \ldots, n$ such that $f(xy) = \sum h_i(x)k_i(y)$ for all $x, y \in G$. Then $f(y^{-1}x^{-1}) = f'(xy) = \sum h_i(y^{-1})k_i(x^{-1}) = \sum k_i'(x)h_i'(y)$. By (2.4), again, f' is representative.

Definition 2.18 Let G be an analytic group. Then $S : R(G) \to R(G)$ is the function defined by $S(f) = f'$. (S is well-defined by (2.7).)

Lemma 2.19 Let G be an analytic group, let $f \in R(G)$, and suppose $f(xy) = \sum h_i(x)k_i(y) = \sum p_i(x)q_i(y)$ with h_i, k_i, p_i, $q_i \in R(G)$. Then $\sum h_i \otimes k_i = \sum p_i \otimes q_i$ in $R(G) \otimes R(G)$.

Proof. We first claim that if $g, h \in R(G)$, the function $(x, y) \mapsto g(x)h(y)$ is in $R(G \times G)$. To see this, choose G-modules, V, W and elements $v \in V$, $w \in W$, $\phi \in V^*$, $\psi \in W^*$ such that $g = f_{\phi,v}$ and $h = f_{\psi,w}$. Then $V \otimes W$ is a $G \times G$-module under the action $(x, y)(u \otimes z) = xu \otimes yz$. We identify $V^* \otimes W^*$ with $(V \otimes W)^*$. Then $g(x)h(y) = \phi(xv)\psi(yw) = (\phi \otimes \psi)((x, y)(v \otimes w)) = f_{\phi \otimes \psi, v \otimes w}(x, y)$, so our function is representative. We thus have a map $r : R(G) \otimes R(G) \to R(G \times G)$ such that $r(g \otimes h)(x, y) = g(x)h(y)$. We next claim that r is injective. Suppose $r(\sum s_i \otimes t_i) = 0$. Let $L = \langle s_1, \ldots, s_n \rangle$ and $R = \langle t_1, \ldots, t_n \rangle$. We can choose bases $\{a_1, \ldots, a_l\}$, $\{b_1, \ldots, b_m\}$ of L, R and elements $x_1, \ldots, x_l$, $y_1, \ldots, y_m$ of G such that $a_i(x_j) = \delta_{ij}$ and $b_i(y_j) = \delta_{ij}$. If $r(\sum \alpha_{ij} a_i \otimes b_j) = 0$, $0 = r(\sum \alpha_{ij} a_i \otimes b_j)(x_c, y_d) = \sum \alpha_{ij} a_i(x_c)b_j(y_d) = \alpha_{cd}$ so $r : L \otimes R \to R(G \times G)$ is injective, so $\sum s_i \otimes t_i = 0$. Finally, in the notation of the hypotheses, $r(\sum h_i \otimes k_i)(x, y) = f(xy) = r(\sum p_i \otimes q_i)(x, y)$ for all $(x, y) \in G \times G$, so the injectivity of r implies $\sum h_i \otimes k_i = \sum p_i \otimes q_i$.

Definition 2.20 Let G be an analytic group. Then $\Delta : R(G) \to R(G) \otimes R(G)$ is the function defined by $\Delta(f) = \sum h_i \otimes k_i$ if $f(xy) = \sum h_i(x)k_i(y)$ for all x, y in G, (Δ is well-defined by (2.19)). Also, $\varepsilon : R(G) \to \mathbb{C}$ is the function defined by $\varepsilon(f) = f(e)$.

Theorem 2.21 Let G be an analytic group. Then R(G) is a Hopf algebra over $\mathbb{C}$ with comultiplication Δ, counit ε, and antipode S.

Proof. We have to verify that ε and Δ are algebra homomorphisms and that three diagrams commute. To express these latter compactly, it is convenient to introduce the map $q : R(G) \to R(G)$ where $q(f)(x) = f(e)$. Then the diagrams become the functional equations:

  (i) (coassociativity) $(1 \otimes \Delta)\Delta = (\Delta \otimes 1)\Delta$;
  (ii) (coidentity) $(1 \otimes q)\Delta = 1 = (q \otimes 1)\Delta$;
  (iii) (coinverses) $(1 \otimes S)\Delta = q = (S \otimes 1)\Delta$.

All these verifications are straightforward from (2.18) and (2.20) and left to the reader.

We can use the Hopf algebra structure of $R(G)$ to algebraically express the left and right actions of G on $R(G)$. We begin with the observation that, by (2.4)(d), if $f \in R(G)$ then $\Delta(f) \in {}_l[f] \otimes [f]_r$. Now if V is a left stable (respectively, right stable) subspace of V and $f \in V$, then ${}_l[f] \subseteq V$ (respectively, $[f]_r \subseteq V$), so $\Delta(V) \subseteq V \otimes R(G)$ (respectively, $\Delta(V) \subseteq R(G) \otimes V$). This property is characteristic, as the following shows.

Proposition 2.22 Let G be an analytic group and V a subspace of $R(G)$. Then V is left stable (respectively, right stable) if and only if $\Delta(V) \subseteq V \otimes R(G)$ (respectively, $\Delta(V) \subseteq R(G) \otimes V$).

Proof. We have just observed the 'only if' part. So assume $\Delta(V) \subseteq V \otimes R(G)$, and let $f \in V$. Then $\Delta(f) = \sum h_i \otimes k_i$, where $h_i \in V$. Thus, $x \cdot f = \sum k_i(x)h_i$ is in V. A similar argument applies on the right.

Now suppose V is a left stable subspace of $R(G)$ and let $w = \Delta|V$. Then elementary calculations show that $(1 \otimes \varepsilon)w$ is the canonical isomorphism $V \to V \otimes \mathbb{C}$ and $(w \otimes 1)w = (1 \otimes \Delta)w$. These are precisely the conditions that make V a right $R(G)$-comodule under the action $w : V \to V \otimes R(G)$.

Definition 2.23 Let G be an analytic group. A *right comodule* for $R(G)$ is a vector space M and a $\mathbb{C}$-linear map $w : M \to M \otimes R(G)$ such that $(1 \otimes \varepsilon)w(m) = m \otimes 1$ for all m in M and $(w \otimes 1)w = (1 \otimes \Delta)w$. We refer to w as the right coaction of $R(G)$ on M.

We are going to show that G-modules are the same thing as finite-dimensional right $R(G)$-comodules.

Proposition 2.24 Let G be an analytic group and let V be a finite-dimensional right $R(G)$-comodule with coaction w. Then V is a G-module where $g(v) = \sum f_i(g)v_i$ if $g \in G$, $v \in V$, and $w(v) = \sum v_i \otimes f_i$.

Proof. Since $(1 \otimes \varepsilon)w(v) = v \otimes 1$ for all v in V, if $w(v) = \sum v_i \otimes f_i$, then $\sum v_i \otimes f_i(e) = v \otimes 1$, or $v = \sum f_i(e)v_i$, so $e(v) = v$. Now suppose $w(v) = \sum v_i \otimes f_i$, $w(v_i) = \sum v_{i,j} \otimes f_{i,j}$ and $\Delta(f_i) = \sum h_{i,j} \otimes k_{i,j}$. Then the equation $(w \otimes 1)w(v) = (1 \otimes \Delta)w(v)$ becomes $\sum_i (\sum_j v_{i,j} \otimes f_{i,j}) \otimes f_i = \sum_i (v_i \otimes \sum h_{i,j} \otimes k_{i,j})$. If x, y are in G, we can map $V \otimes R(G) \otimes R(G)$ to V by $v \otimes f \otimes g \mapsto f(x)g(y)v$. Then this equation becomes $\sum_i (\sum_j f_{i,j}(x)v_{i,j})f_i(y) = \sum_i (\sum_j h_{i,j}(x)k_{i,j}(y))v_i$. We want to analyze this equation. By definition, $y(v) = \sum f_i(y)v_i$ so $x(y(v)) = \sum f_i(y)x(v_i)$, and $x(v_i) = \sum f_{i,j}(x)v_{i,j}$, so the left-hand side is $x(y(v))$. For the right-hand side, $\sum_j h_{i,j}(x)k_{i,j}(y) = f_i(xy)$, so the right-hand side is $(xy)(v)$. Thus, $x(y(v)) = xy(v)$ and we do have an action. Since each element of $R(G)$ is analytic, we have that V is actually a G-module.

To obtain the converse of (2.24) we need two additional observations.

Lemma 2.25 Let G be an analytic group and let W be a left stable subspace of $R(G)$. Then the G-module structure on W obtained from the $R(G)$-comodule structure of (2.22) is $x(h) = x \cdot h$ for $x \in G$ and $h \in W$.

Proof. The G-module structure from (2.24) is given by $x(h) = \sum q_i(x)p_i$ if $\Delta(h) = \sum p_i \otimes q_i$. Since $\sum p_i(y)q_i(x) = h(yx) = (x \cdot h)(y)$, $x \cdot h = \sum q_i(x)p_i$ also.

Lemma 2.26 Let G be an analytic group, V a finite-dimensional right $R(G)$-comodule, and let W be a G-submodule of V in the action of (2.24). Then $w(W) \subseteq W \otimes R(G)$, where w is the coaction of $R(G)$ on V.

Proof. Let $v \in W$ and suppose $w(v) = \sum v_i \otimes f_i$. We can assume by (2.1) that there are $x_1, \ldots, x_n \in G$ with $f_i(x_j) = \delta_{ij}$. (First take $f_1, \ldots, f_n$ linearly independent and then change to a basis of $\langle f_1, \ldots, f_n \rangle$ of the appropriate form by (2.1).) Then $x_j(v) = \sum f_i(x_j)v_i = v_j$, so $v_j = x_j(v) \in W$ and $w(v) \in W \otimes R(G)$.

We can now put a comodule structure on all G-modules.

Proposition 2.27 Let G be an analytic group and let V be a G-module. Then there is an $R(G)$-coaction w on V such that $w(v) = \sum v_i \otimes f_i$ if $g(v) = \sum f_i(g)v_i$ for all g in G.

Proof. By (2.14), we have a G-module injection $V \to \prod W_i$ where each W_i is a finite-dimensional left stable subspace of $R(G)$. Let $W = \prod W_i$. Define $d: W \to W \otimes R(G)$ by $d(f_1, \ldots, f_n) = (\Delta(f_1), \ldots, \Delta(f_n))$. By (2.22), W is a comodule with coaction d. By (2.26), $w = d|V$ is a coaction of $R(G)$ on V. We need only verify the formula for the coaction on elements of W. If $f = (f_1, \ldots, f_n)$ is in W, and $x \in G$, then $x(f) = (x(f_1), \ldots, x(f_n))$. By (2.25), $x(f_i) = \sum q_{ij}(x)p_{ij}$ if $\Delta(f_i) = \sum p_{ij} \otimes q_{ij}$. Thus, $d(f) = \sum_{i,j} (0, \ldots, 0, p_{ij}, 0, \ldots, 0) \otimes q_{ij}$ and $x(f) = \sum_{i,j} q_{ij}(x)(0, \ldots, 0, p_{ij}, 0, \ldots, 0)$. To complete the proof, we need to know that if $v_i \in V$ and $f_i \in R(G)$, $i = 1, \ldots, n$, then $\sum v_i \otimes f_i$ is determined by the function $G \to V$ which sends x to $\sum f_i(x)v_i$. Since if two elements of $V \otimes R(G)$ give the same function their difference gives the zero function, it will be enough to show that if $\sum v_i \otimes f_i$ gives the zero function, $\sum v_i \otimes f_i = 0$. Let $k_1, \ldots, k_m$ be a basis of $\langle f_1, \ldots, f_n \rangle$ such that there are $x_1, \ldots, x_m$ in G with $k_i(x_j) = \delta_{ij}$ (2.1). Then $\sum v_i \otimes f_i = \sum u_i \otimes k_i$ for suitable $u_i \in V$, and $u_j = \sum k_i(x_j)u_i = \sum f_i(x_j)v_i = 0$ for all j, so $\sum v_i \otimes f_i = 0$.

Combining (2.24) and (2.27) yields a correspondence between G-modules and $R(G)$-comodules. This correspondence will actually be an equivalence of categories, once we recall the definition of morphisms of comodules.

Definition 2.28 Let G be an analytic group and let V and V' be $R(G)$-comodules with coactions w and w'. A linear transformation $f: V \to V'$ is a *comodule morphism* if $(f \otimes 1)w = w'f$.

$R(G)$-comodules and their morphisms then form a category. We denote the full subcategory of finite-dimensional comodules by Co-Mod$_{\text{fd}}(R(G))$.

Theorem 2.29 Let G be an analytic group. The categories Mod(G) and Co-Mod$_{\text{fd}}(R(G))$ are equivalent.

Proof. The objects of the two categories are the same and the module and comodule structures correspond by (2.24) and (2.27), the correspondence clearly being functorial.

Because of (2.29), we know a good deal about the category Co-Mod$_{\text{fd}}(R(G))$: it is abelian, it has a tensor product structure, and we can generate it from subcomodules of $R(G)$, since similar statements apply to Mod(G). Details will not be pursued here, as we will not study comodules in any depth. We merely observe that, in principle, (2.29) says that the category Mod(G) can be recovered from the Hopf algebra structure on $R(G)$. Thus, the analytic G-modules are algebraically obtained from $R(G)$.

As we saw in Chapter 1, a G-module structure on the vector space V corresponds to an analytic representation $\rho : G \to \text{GL}(V)$. Thus, such representations are obtained from the Hopf algebra $R(G)$. We want to make this explicit.

We fix a representation $\rho : G \to \text{GL}(V)$. Let $A = \mathbb{C}[\text{GL}(V)]$, the affine coordinate ring of the algebraic group $\text{GL}(V)$; A is a Hopf algebra with comultiplication $\mu^*(f) = \sum h_i \otimes k_i$ if $f(xy) = \sum h_i(x)g_i(y)$, counit $e^*(f) = f(e)$ and antipode $i^*(f)(x) = f(x^{-1})$, if $f \in A$. We can describe A as an algebra as follows: fix a basis $v_1, \ldots, v_n$ of V and let $\phi_1, \ldots, \phi_n$ be the dual basis. Let $m_{ij} \in A$ be given by $m_{ij}(x) = \phi_i(xv_j)$. (The matrix of x in the chosen basis is then $[m_{ij}(x)]$.) Let $d \in A$ be given by $d(x) = \det(x)$. (So d is the determinant of the matrix $[m_{ij}]$.) Then $A = \mathbb{C}[m_{ij} \mid 1 \leq i, j \leq n][d^{-1}]$, and the m_{ij} are algebraically independent. Let $B = \mathbb{C}[m_{ij} \mid 1 \leq 1, j \leq n]$. Define $l_{ij} : G \to \mathbb{C}$ by $l_{ij}(x) = m_{ij}(\rho(x)) = \phi_i(\rho(x)v_j)$. By (2.7), $l_{ij} \in R(G)$. Define $h : B \to R(G)$ by $h(m_{ij}) = l_{ij}$. We want to extend the $\mathbb{C}$-algebra homomorphism h to A, i.e., we want to show that $h(d)$ is a unit of $R(G)$. Now $h(d)(x) = \det(\rho(x))$, and, in the notation of (2.16), $(h(d)')(x) = h(d)(x^{-1}) =$

$\det(\rho(x^{-1})) = (\det(\rho(x))^{-1} = (h(d)(x))^{-1}$. Thus, $h(d)$ is a unit of $R(G)$ with $h(d)^{-1} = h(d)'$. Hence, h extends to a $\mathbb{C}$-algebra homomorphism $A \to R(G)$ which we also call h. Let us observe that the homomorphism does not depend on the choice of basis of V: if $v \in V$ and $\phi \in V^*$, let $m_{\phi,v}$ in A be given by $m_{\phi,v}(x) = \phi(xv)$. Then it is easy to see that $h(m_{\phi,v})(x) = \phi(\rho(x)v)$, so h is independent of the choice of basis. In fact, we have that $h(f)(x) = f(\rho(x))$ for every f in A since this formula holds when f is one of the generators m_{ij}.

We now show that h is also a Hopf algebra homomorphism: $\varepsilon h(f) = h(f)(e) = f(\rho(e)) = e^*(f)$ for f in A, so $\varepsilon h = e^*$, and $S(h(f))(x) = h(f)(x^{-1}) = f(\rho(x)^{-1}) = h(i^*f)(x)$ so $Sh = hi^*$, and if $\mu^*(f) = \sum h_i \otimes k_i$, then $(h \otimes h)(\mu^*(f)) = \sum h(h_i) \otimes h(k_i)$, and $\sum h(h_i)(x)h(k_i)(y) = \sum h_i(\rho(x))k_i(\rho(y)) = f(\rho(x)\rho(y)) = h(f)(xy)$, so $\Delta h = (h \otimes h)\mu^*$.

Conversely, suppose we are given a Hopf algebra homomorphism $k: A = \mathbb{C}[\mathrm{GL}(V)] \to R(G)$. If $x \in G$, then $A \to \mathbb{C}$ by $f \mapsto k(f)(x)$ is a $\mathbb{C}$-algebra homomorphism, and, hence, defines an element $\rho(x)$ of $\mathrm{GL}(V)$ such that $f(\rho(x)) = k(f)(x)$ for all f in A. If $f \in A$ and $\mu^*f = \sum p_i \otimes q_i$, then $f(\rho(x)\rho(y)) = \sum p_i(\rho(x))q_i(\rho(y)) = \sum k(p_i)(x)k(q_i)(y) = k(f)(xy) = f(\rho(xy))$, the third equality coming from the formula $\Delta k = (k \otimes k)\mu^*$. Thus, $\rho(x)\rho(y) = \rho(xy)$ and ρ is a group homomorphism. If $\phi \in V^*$ and $v \in V$, $m_{\phi,v}(\rho(x)) = k(m_{\phi,v})(x)$ is an analytic function of x, so $\rho: G \to \mathrm{GL}(V)$ is an analytic homomorphism, and, hence, a representation of G. It is clear that the construction of this ρ from k is inverse to the construction of h from ρ above. So we have established the following theorem.

Theorem 2.30 Let G be an analytic group and V a finite-dimensional vector space. Representations $\rho: G \to \mathrm{GL}(V)$ correspond to Hopf algebra homomorphisms $h: \mathbb{C}[\mathrm{GL}(V)] \to R(G)$, where $h(f)(x) = f(\rho(x))$ for all f in $\mathbb{C}[\mathrm{GL}(V)]$.

To summarize: the analytic representation theory of G is completely determined by the Hopf algebra $R(G)$.

If $\rho: G \to \mathrm{GL}(V)$ is a representation and $h: \mathbb{C}[\mathrm{GL}(V)] \to R(G)$ is the corresponding Hopf algebra homomorphism, then the image

B of h is a Hopf subalgebra of $R(G)$ finitely-generated over $\mathbb{C}$ (and without nilpotent elements), and the Hopf algebra surjection $\mathbb{C}[GL(V)] \to B$ gives an injection of algebraic groups $H \to GL(V)$, where H is the affine algebraic group with coordinate ring B. For reasons that will become clear in Chapter 3, we want to determine the image of H in $GL(V)$.

Proposition 2.31 Let G be an analytic group, V a finite-dimensional vector space and $\rho: G \to GL(V)$ a representation. Let $h: \mathbb{C}[GL(V)] \to R(G)$ be the corresponding Hopf algebra homomorphism, and let $\bar{G} = \{x \in GL(V) \mid f(x) = 0 \text{ for all } f \in \text{Ker}(h)\}$. Then $\bar{G}$ is the Zariski-closure of $\rho(G)$ in $GL(V)$.

Proof. (If $B = \text{Image}(h)$, $H = $ the algebraic group with coordinate ring B, and $H \to GL(V)$ is the algebraic group homomorphism arising from $\mathbb{C}[GL(V)] \to B$, then $\bar{G}$ is the image of H, so (2.31) does solve the determination problem just mentioned.) By (2.30), we have $h(f)(x) = f(\rho(x))$ for all f in $A = \mathbb{C}[GL(V)]$, so $\rho(G) \subseteq \bar{G}$. Since $\bar{G}$ is Zariski-closed in $GL(V)$, we need only show that $\rho(G)$ is Zariski-dense in $\bar{G}$, i.e., if $f \in A$ and $f|\rho(G) = 0$, then $f|\bar{G} = 0$. If $f|\rho(G) = 0$, then for all x in G, $0 = f(\rho(x)) = h(f)(x)$, so $f \in \text{Ker}(h)$ and $f|\bar{G} = 0$.

We now turn to some groups associated with $R(G)$. These will be shown to be isomorphic, and ultimately isomorphic to the group of tensor automorphisms of $\text{Mod}(G)$.

Definition 2.32 Let G be an analytic group. Then $\mathscr{G}(G)$ is the set of all $\mathbb{C}$-algebra homomorphisms from $R(G)$ to $\mathbb{C}$. If $x \in G$, $\tau(x) \in \mathscr{G}(G)$ is defined by $\tau(x)(f) = f(x)$.

We can define some operations on $\mathscr{G}(G)$ from the Hopf algebra structure on $R(G)$: (i) if $a, b \in \mathscr{G}(G)$, let $a * b \in \mathscr{G}(G)$ be given by $(a * b)(f) = \sum a(h_i)b(k_i)$ if $\Delta(f) = \sum h_i \otimes k_i$; (ii) if $a \in \mathscr{G}(G)$, let $a' \in \mathscr{G}(G)$ be given by $a'(f) = a(f')$.

Proposition 2.33 Let G be an analytic group. Then $\mathscr{G}(G)$ is a group with multiplication $*$, identity $\tau(e)$ and inverse $(\ )'$. Moreover, $\tau: G \to \mathscr{G}(G)$ is a group homomorphism.

Proof. That $\mathscr{G}(G)$ is a group under the given operations is a standard Hopf algebra exercise, which we leave to the reader. We will verify that τ is a homomorphism. Let $x, y \in G$, let $f \in R(G)$, and assume $\Delta(f) = \sum h_i \otimes k_i$. Then $(\tau(x) * \tau(y))(f) = \sum \tau(x)(h_i)\tau(y)(k_i) = \sum h_i(x)k_i(y) = f(xy) = \tau(xy)(f)$, so $\tau(x)\tau(y) = \tau(xy)$.

There is another group associated with $R(G)$ which is isomorphic to $\mathscr{G}(G)$ but sometimes more convenient to deal with.

Definition 2.34 Let G be an analytic group. Propaut($R(G)$) is the set of all $\mathbb{C}$-algebra automorphisms α of $R(G)$ such that $\Delta\alpha = (1 \otimes \alpha)\Delta$. Elements of Propaut($R(G)$) are called *proper automorphisms*.

It is easy to see that Propaut($R(G)$) is a group under composition of automorphisms. The existence of nonidentity proper automorphisms is shown in the following.

Proposition 2.35 Let G be an analytic group. If $x \in G$, $l_x : R(G) \to R(G)$ given by $l_x(f) = x \cdot f$ is a proper automorphism, and $\gamma : G \to$ Propaut($R(G)$) by $\gamma(x) = l_x$ is a group homomorphism.

Proof. Let $f \in R(G)$ and let $\Delta(f) = \sum h_i \otimes k_i$. Then $(1 \otimes l_x)\Delta(f) = \sum h_i \otimes x \cdot k_i$. If $y, z \in G$, $\sum h_i(y)(x \cdot k_i)(z) = \sum h_i(y)k_i(zx) = f(yzx) = (x \cdot f)(yz)$, so $\Delta(l_x f) = \sum h_i \otimes x \cdot k_i = (1 \otimes l_x)\Delta(f)$, and l_x is a proper automorphism. The formula $l_{xy} = l_x l_y$, which follows from $xy \cdot f = x \cdot (y \cdot f)$, shows γ is a homomorphism.

Theorem 2.36 Let G be an analytic group. Then $h :$ Propaut($R(G)$) $\to \mathscr{G}(G)$ by $h(\alpha) = \varepsilon\alpha$ is a group isomorphism.

Proof. Let α, β be in Propaut($R(G)$). Then $(\varepsilon\alpha * \varepsilon\beta)(f) = \sum \alpha(h_i)(e)\beta(k_i)(e)$ if $\Delta(f) = \sum h_i \otimes k_i$. Let $\Delta(\beta f) = \sum p_i \otimes q_i$. Since $\Delta\beta = (1 \otimes \beta)\Delta$, $\Delta(\beta(f)) = \sum h_i \otimes \beta(k_i)$. Since $\beta(f)(x) = \beta(f)(xe) = \sum p_i(x)q_i(e)$, $\beta(f) = \sum p_i q_i(e) = \sum h_i \beta(k_i)(e)$. Thus, $\alpha\beta(f) = \sum \alpha(h_i)\beta(k_i)(e)$ so $(\varepsilon\alpha\beta)(f) = \alpha\beta(f)(e) = \sum \alpha(h_i)(e)\beta(k_i)(e) = (\varepsilon\alpha * \varepsilon\beta)(f)$ and h is a homomorphism. Now let $a \in \mathscr{G}(G)$, and $f \in R(G)$. Define $\alpha(f) = \sum h_i a(k_i)$ if $\Delta(f) = \sum h_i \otimes k_i$, so α is the

composite $R(G) \to^{\Delta} R(G) \otimes R(G) \to^{1 \otimes a} R(G) \otimes \mathbb{C} \to R(G)$,
which we will write as $\alpha = (1 \otimes a)\Delta$. Let $b \in \mathcal{G}(G)$, let $\beta = (1 \otimes b)\Delta$,
and let $\gamma = (1 \otimes (a * b))\Delta$. Let $f \in R(G)$. We claim that $\alpha\beta(f) =$
$\gamma(f)$. For suppose $\Delta(f) = \sum h_i \otimes k_i$ and $\Delta(h_i) = \sum p_{ij} \otimes q_{ij}$. Then
$\alpha\beta(f) = \sum \alpha(h_i)b(k_i) = \sum p_{ij}a(q_{ij})b(k_i)$. If $\Delta(k_i) = \sum r_{ij} \otimes s_{ij}$, then
$\gamma(f) = \sum h_i(a * b)(k_i) = \sum h_i a(r_{ij})b(s_{ij})$. Thus, $(\alpha\beta)(f) = \gamma(f)$ if $\sum p_{ij} \otimes$
$q_{ij} \otimes k_i = \sum h_i \otimes r_{ij} \otimes s_{ij}$. But the former is $(\Delta \otimes 1)\Delta(f)$ and the
latter is $(1 \otimes \Delta)\Delta(f)$, and these are equal. If $b = a'$, then $a * b =$
$a * a' = \tau(e)$, and $(1 \otimes \tau(e))\Delta(f) = \sum h_i k_i(e) = f$, so α is invertible,
and, hence, an algebra automorphism of $R(G)$. In fact, it is a
proper automorphism: $(1 \otimes \alpha)\Delta(f) = \sum h_i \otimes \alpha(k_i)$, and if $x, y \in G$,
$\sum h_i(x)\alpha(k_i)(y) = \sum h_i(x)r_{ij}(y)a(s_{ij})$. Since $\sum h_i \otimes r_{ij} \otimes s_{ij} = \sum p_{ij} \otimes$
$q_{ij} \otimes k_i$, this latter sum is also $\sum p_{ij}(x)q_{ij}(y)a(k_i) = \sum h_i(xy)a(k_i) =$
$\alpha(f)(xy)$, so $\Delta\alpha(f) = \sum h_i \otimes \alpha(k_i)$ and $(1 \otimes \alpha)\Delta = \Delta\alpha$. Thus,
$\alpha \in \text{Propaut}(R(G))$, and $h(\alpha)(f) = \alpha(f)(e) = \sum h_i(e)a(k_i) =$
$a(\sum h_i(e)k_i) = a(f)$ so $h(\alpha) = a$. Thus, h is surjective. Finally,
$(1 \otimes h(\alpha))\Delta = (1 \otimes \varepsilon)(1 \otimes \alpha)\Delta = (1 \otimes \varepsilon)\Delta\alpha = \alpha$, so h is also injective and, hence, an isomorphism.

We are now going to show that the groups $\text{Propaut}(R(G))$ and
$\mathcal{G}(G)$ are isomorphic to the group $\text{Aut}_{\otimes}(\text{Mod}(G))$ introduced in
Chapter 1 (see (1.7) and (1.8)).

For convenience, we fix $\alpha \in \text{Aut}_{\otimes}(\text{Mod}(G))$. If V is a G-module
and $v \in V$, we will write $\alpha(v)$ for $\alpha_V(v)$ when no confusion will
arise. We define a map $a : R(G) \to R(G)$ from α by $a(f) = \alpha(f)$,
where we regard f as an element of the G-module $_1[f]$. We are
going to show that a is a proper automorphism of $R(G)$. First, we
remark that if V, W are finite-dimensional left stable subspaces of
$R(G)$ with $V \subseteq W$ then $\alpha_W|V = \alpha_V$, so if $f \in V$ then $a(f) = \alpha_V(f)$.
It follows that a is a linear endomorphism of $R(G)$, since if f,
$g \in R(G)$ and $V = _1[f] + _1[g]$, then $\alpha_V(f+g) = \alpha_V(f) + \alpha_V(g)$, so
$a(f+g) = a(f) + a(g)$. Further, if $U = _1[f]$ and $W = _1[g]$, and
$h : V \otimes W \to VW$ ($\subseteq R(G)$) is the product map, then h is a G-
morphism so $\alpha_{VW}h = h(\alpha_V \otimes \alpha_W)$, and, hence, $a(fg) = a(f)a(g)$
(apply the functional equation to $f \otimes g$). Since on $_1[f]$ a is a linear
automorphism, a is an algebra automorphism of $R(G)$. Now let
$f \in R(G)$ and let V be a finite-dimensional left and right stable

subspace of V containing f. For each x in G, the function $r_x\colon V \to V$ given by $r_x(g) = g \cdot x$ is a G-module endomorphism. So $\alpha_V r_x = r_x \alpha_V$, or $a(f \cdot x) = a(f) \cdot x$. Let $\Delta(f) = \sum h_i \otimes k_i$. Then $f \cdot x = \sum h_i(x)k_i$ so $a(f) \cdot x = a(f \cdot x) = \sum h_i(x)a(k_i)$. Evaluating both sides at $y \in G$, we find that $a(f)(xy) = \sum h_i(x)a(k_i)(y)$, or $(\Delta a)(f) = (1 \otimes a)\Delta(f)$. This shows that a is a proper automorphism of $R(G)$. If we now write $a = a(\alpha)$, and let $\beta \in \mathrm{Aut}_\otimes(\mathrm{Mod}(G))$, then it is immediate from the definitions that $a(\alpha\beta) = a(\alpha)a(\beta)$, so that $a\colon \mathrm{Aut}_\otimes(\mathrm{Mod}(G)) \to \mathrm{Propaut}(R(G))$ is a group homomorphism. In fact, more is true.

Theorem 2.37 Let G be an analytic group and let $a\colon \mathrm{Aut}_\otimes(\mathrm{Mod}(G)) \to \mathrm{Propaut}(R(G))$ be given by $a(\alpha)(f) = \alpha_V^{\cdot}(f)$, where $V = {}_1[f]$. Then a is a group isomorphism.

Proof. We have already seen that a is a well-defined group homomorphism. We will show it is an isomorphism by essentially constructing an inverse. We will do this via the isomorphism of (2.36). Let $g \in \mathscr{G}(G)$, and let V be a G-module. Then by (2.27), V is also a right $R(G)$-comodule with coaction w such that $w(v) = \sum v_i \otimes f_i$ if $x(v) = \sum f_i(x)v_i$ for all x in G. Let $\alpha_V\colon V \to V$ be the composite $V \to^w V \otimes R(G) \to^{1 \otimes g} V \otimes \mathbb{C} \to V$, so $\alpha_V(v) = \sum g(f_i)v_i$ if $w(v) = \sum v_i \otimes f_i$. Now suppose V' is also a G-module, viewed as a right $R(G)$-comodule with coaction w' and let $f\colon V \to V'$ be a G-module homomorphism. Then f is also an $R(G)$-comodule morphism, so $(f \otimes 1)(1 \otimes g)w = (1 \otimes g)(f \otimes 1)w = (1 \otimes g)w'f$, and it follows that $f\alpha_V = \alpha_{V'}f$. Finally, let V, U be G-modules. We want to compare $\alpha_V \otimes \alpha_U$ with $\alpha_{V \otimes U}$. We view V, U and $V \otimes U$ as $R(G)$-comodules, denoting all three coactions by w. Suppose $u \in U$, $v \in V$, $w(u) = \sum u_i \otimes f_i$ and $w(v) = \sum v_i \otimes g_i$. Then, if $x \in G, x(v \otimes u) = x(v) \otimes x(u) = (\sum g_i(x)v_i) \otimes (\sum f_i(x)u_j) = \sum (g_if_j)$ $(x) \times (v_i \otimes u_j)$, so $w(v \otimes u) = \sum v_i \otimes u_j \otimes g_if_j$. Then, $\alpha_{V \otimes U}(v \otimes u) = \sum h(g_if_j)(v_i \otimes u_j) = \sum h(g_i)h(f_j)(v_i \otimes u_j) = (\sum h(g_i)v_i) \otimes (\sum h(f_j)v_j)$ $= \alpha_V(v) \otimes \alpha_U(u)$, so $\alpha_{V \otimes U} = \alpha_V \otimes \alpha_U$. Thus, α is what might be termed a 'tensor endomorphism' of $\mathrm{Mod}(G)$. We write $\alpha = \alpha(g)$ to indicate the dependence of α on $g \in \mathscr{G}(G)$. Now let $g_1, g_2 \in \mathscr{G}(G)$. If we suppress the isomorphism $V \otimes \mathbb{C} \otimes \mathbb{C} \to V$, then $\alpha(g_1 * g_2)_V = (1 \otimes g_1 \otimes g_2)(1 \otimes \Delta)w = (1 \otimes g_1 \otimes g_2)(w \otimes 1)w = (1 \otimes g_1 \otimes 1) \times$

$(1 \otimes 1 \otimes g_2)(w \otimes 1)w = ((1 \otimes g_1)w \otimes 1)(1 \otimes g_2)w = \alpha(g_1)_v \alpha(g_2)_v$.
It follows from the definition of comodule that $\alpha(\varepsilon)$ is the identity, and ε is the identity of $\mathcal{G}(G)$, so $\alpha(g)$ is invertible with inverse $\alpha(g^{-1})$. Thus, $\alpha(g)$ is actually a tensor automorphism of $\mathrm{Mod}(G)$, for $g \in \mathcal{G}(G)$. In (2.36) we have an isomorphism $h: \mathrm{Propaut}(R(G)) \to \mathcal{G}(G)$ by $h(c) = \varepsilon c$. Then for $f \in R(G)$, $a(\alpha(h(c))(f) = \alpha(h(c))_v(f)$, where V is any left stable finite-dimensional subspace of $R(G)$ containing f. Now the coaction on V is given by $\Delta | V$ by (2.22), and if $\Delta(f) = \sum p_i \otimes q_i$, then $\alpha(h(c))_v(f) = \sum h(c)(q_i)p_i = \sum c(q_i)(e)p_i = \sum p_i c(q_i)(e) = (1 \otimes \varepsilon)(1 \otimes c)\Delta(f) = (1 \otimes \varepsilon)\Delta(c(f)) = c(f)$, so $a(\alpha(h(c)) = c$. We have now shown that a is onto. Suppose β lies in the kernel of a. Then $\beta_V = a(\beta)_V = 1_V$ for all left stable finite-dimensional subspaces V of $R(G)$, so if W is any G-module and $p: W \to V$ any G-module homomorphism, $p = p\beta_W$. By (2.14), we know there are left stable finite-dimensional subspaces $V_1, \ldots, V_n$ of $R(G)$ and G-module homomorphisms $p_i: W \to V_i$ such that $q = (p_1, \ldots, p_n): W \to \prod V_i$ is a G-module injection. Thus, $q\beta_W = (p_1\beta_W, \ldots, p_n\beta_W) = (p_1, \ldots, p_n) = q$, so β_W is the identity, and, hence, β is the identity. Thus, a has a trivial kernel and, hence, is an isomorphism.

Theorem 2.37 shows that the group $\mathrm{Propaut}(R(G))$ and its alter ego $\mathcal{G}(G)$ are determined from the category $\mathrm{Mod}(G)$. This suggests that the Hopf algebra $R(G)$ can be recovered from the category $\mathrm{Mod}(G)$, and we will show that this is, in fact, the case. First, however, a few comments on (2.37): note that the proof used, in an essential way, the isomorphism between $\mathrm{Propaut}(R(G))$ and $\mathcal{G}(G)$ and the equivalence of G-modules and $R(G)$-comodules. This is not accidental: many of the constructions we need to discuss are simple from one point of view and obscure from another. Our various isomorphisms and equivalences give us a certain amount of freedom in choosing a point of view which, as the proof of (2.37) shows, can be quite helpful.

We recall that we have homomorphisms $\sigma: G \to \mathrm{Aut}_{\otimes}(\mathrm{Mod}(G))$ (1.9), $\tau: G \to \mathcal{G}(G)$ (2.32), and $\gamma: G \to \mathrm{Propaut}\, R(G)$ (2.35). These are related by the isomorphisms $h: \mathrm{Propaut}(R(G)) \to \mathcal{G}(G)$ (2.36) and $a: \mathrm{Aut}_{\otimes}(\mathrm{Mod}(G)) \to \mathrm{Propaut}(R(G))$ (2.37) as follows: $a\sigma = \gamma$ and $h\gamma = \tau$. To see the first equation, let $x \in G$ and $f \in R(G)$. Then

$a(\sigma(x))(f) = \sigma(x)(f) = x \cdot f = l_x(f) = \gamma(x)(f)$, so $a\sigma(x) = \gamma(x)$. To
see the second, let $x \in G$ and $f \in R(G)$. Then $h\gamma(x)(f) = (\varepsilon l_x)(f) = l_x(f)(e) = (x \cdot f)(e) = f(x) = \tau(x)(f)$, so $h\gamma(x) = \tau(x)$. These observations will be used below to put $\mathscr{G}(G)$-actions on G-modules.

We now observe that G-modules are also $\mathrm{Aut}_\otimes(\mathrm{Mod}(G))$-modules.

Definition 2.38 Let G be an analytic group and let V be a
G-module. If $\alpha \in \mathrm{Aut}_\otimes(\mathrm{Mod}(G))$ and $v \in V$, then $\alpha(v) = \alpha_V(v)$. This
gives an action of $\mathrm{Aut}_\otimes(\mathrm{Mod}(G))$ on V; if W is also a G-module
and $f: V \to W$ a G-module homomorphism, then $f(\alpha(v)) = \alpha(f(v))$.
(Verifications are left to the reader.)

Thus $\mathrm{Mod}(G)$ is also a category of vector spaces with
$\mathrm{Aut}_\otimes(\mathrm{Mod}(G))$ action. To see which vector spaces occur, we need
the notion of a representative function on $\mathrm{Aut}_\otimes(\mathrm{Mod}(G))$.

Definition 2.39 Let G be an analytic group. A function
$f: \mathrm{Aut}_\otimes(\mathrm{Mod}(G)) \to \mathbb{C}$ is *representative* if there is a G-module V,
$v \in V$, and $\phi \in V^*$ such that $f(\alpha) = \phi(\alpha(v))$ for all α.

In the notations of (2.39), (2.8) and (1.9), $f\sigma = f_{\phi,v}$.

Notation 2.40 Let G be an analytic group; $R(\mathrm{Mod}(G))$ is the set
of all representative functions on $\mathrm{Aut}_\otimes(\mathrm{Mod}(G))$. (Note that the
notation is chosen to emphasize the fact that this set only depends
on the category $\mathrm{Mod}(G)$.)

Lemma 2.41 Let G be an analytic group. Then $\sigma^*: R(\mathrm{Mod}(G)) \to R(G)$ by $(\sigma^* f)(x) = f(\sigma(x))$ is a bijection.

Proof. We have already observed that σ^* is well-defined. By the
remark following (2.8), σ^* is surjective. Suppose $f, g \in R(\mathrm{Mod}(G))$
and $\sigma^* f = \sigma^* g$. There are G-modules V, W, elements $v \in V$, $w \in W$, $\phi \in V^*$ and $\psi \in W^*$ such that $f(\alpha) = \phi(\alpha(v))$ and such that
$f(\alpha) = \phi(\alpha(v))$ and $g(\alpha) = \psi(\alpha(w))$. Let $U = V \oplus W$, $u = (v, w)$,
and define $\chi \in U^*$ by $\chi(a, b) = \phi(a) - \psi(b)$. Then U is a G-module and $\chi(\alpha(u)) = \chi(\alpha v, \alpha w) = \phi(\alpha(v)) - \psi(\alpha(w))$. Define
$h: \mathrm{Aut}_\otimes(\mathrm{Mod}(G)) \to \mathbb{C}$ by $h(\alpha) = \chi(\alpha(u))$. Then $h = f - g$, and

$\sigma^* h(x) = 0$ for all x in G. Let $U' = \langle x(u) \mid x \in G \rangle$. Then U' is a G-submodule of U, and since $\sigma^* h = 0$, $U' \subseteq \mathrm{Ker}(\chi)$. But then $\alpha(U') \subseteq U'$ for all $\alpha \in \mathrm{Aut}_\otimes(\mathrm{Mod}(G))$, so $h(\alpha) = \chi(\alpha(u)) = 0$ for all α, so $h = 0$. Thus, $f = g$.

We can use (2.41) to regard $R(\mathrm{Mod}(G))$ as a Hopf algebra. The algebra structure is just given by pointwise operations on functions. The counit c is given by $c(f) = f(e)$, and the comultiplication d is given by $d(f) = \sum h_i \otimes k_i$ if $\Delta \sigma^*(f) = \sum \sigma^*(h_i) \otimes \sigma^*(k_i)$. This latter equation means that $f(\sigma(xy)) = \sum h_i(\sigma(x)) k_i(\sigma(y))$ for all $x, y \in G$. Fix $y \in G$ and define $g \in R(\mathrm{Mod}(G))$ by $g = \sum k_i(\sigma(y)) h_i$. Then $\sigma^* g(x) = f(\sigma(xy)) = f(\sigma(x)\sigma(y))$. There is a G-module V, $v \in V$, and $\phi \in V^*$ such that $f(\alpha) = \phi(\alpha(v))$. Then $f(\alpha\sigma(y)) = \phi(\alpha(\sigma(y)v))$, so $h(\alpha) = f(\alpha\sigma(y))$ is also representative, and $\sigma^* h(x) = f(\sigma(x)\sigma(y)) = \sigma^* g(x)$, so $h = g$ and $f(\alpha\sigma(y)) = \sum h_i(x) k_i(\sigma(y))$ for all y in G. Now fix $\alpha \in \mathrm{Aut}_\otimes(\mathrm{Mod}(G))$ and define $k : \mathrm{Aut}_\otimes(\mathrm{Mod}(G)) \to \mathbb{C}$ by $k(\beta) = f(\alpha\beta) = \phi((\alpha\beta)(v)) = (\phi\alpha_V) \circ (\beta(v))$. Since $v \in V$ and $\phi\alpha_V \in V^*$, we have $k \in R(\mathrm{Mod}(G))$. For $y \in G$, $(\sigma^* k)(y) = k(\sigma(y)) = f(\alpha\sigma(y)) = \sum h_i(\alpha) k_i(\sigma(y)) = \sum h_i(\alpha) \sigma^* k_i(y) = \sigma^*(\sum h_i(\alpha) k_i)$, so $k = \sum h_i(\alpha) k_i$ and $f(\alpha\beta) = k(\beta) = \sum h_i(\alpha) k_i(\beta)$. If conversely we have $p_i, q_i \in R(\mathrm{Mod}(G))$ with $f(\alpha\beta) = \sum p_i(\alpha) q_i(\beta)$ for all α, β in $\mathrm{Aut}_\otimes(\mathrm{Mod}(G))$, then $(\sigma^* f)(xy) = f(\sigma(x)\sigma(y)) = \sum \sigma^* p_i(x) \sigma^* q_i(y)$ for all x, y in G, so $\Delta(\sigma^* f) = \sum \sigma^* p_i \otimes \sigma^* q_i$, and $df = \sum p_i \otimes q_i$. Hence, $df = \sum h_i \otimes k_i$ if and only if $f(\alpha\beta) = \sum h_i(\alpha) k_i(\beta)$ for all α, β. Let $s : R(\mathrm{Mod}(G)) \to R(\mathrm{Mod}(G))$ be the antipode, so $s(f)(\sigma(x)) = f(\sigma(x)^{-1})$ for all x in G and f in $R(\mathrm{Mod}(G))$. Fix f and let V be a G-module with $v \in V$ and $\phi \in V^*$ such that $f(\alpha) = \phi(\alpha v)$. Let $\psi \in V^{**}$ be evaluation at v. Then V^* is an $\mathrm{Aut}_\otimes(\mathrm{Mod}(G))$ module with $\psi(\alpha(\phi)) = \alpha(\phi)(v) = \phi(\alpha^{-1} v) = f(\alpha^{-1})$, so $\alpha \mapsto f(\alpha^{-1})$ is in $R(\mathrm{Mod}(G))$. Let $g \in R(\mathrm{Mod}(G))$ be $g(\alpha) = f(\alpha^{-1})$. Then $(\sigma^* g)(x) = f(\sigma(x)^{-1}) = s(f)(\sigma(x)) = (\sigma^* s(f))(x)$ for all x in G, so $\sigma^* g = \sigma^* s(f)$ and $s(f) = g$. So we can define the Hopf algebra structure of $R(\mathrm{Mod}(G))$ directly in terms of $\mathrm{Aut}_\otimes(\mathrm{Mod}(G))$.

Theorem 2.42 Let G be an analytic group. Then $R(\mathrm{Mod}(G))$ is a Hopf algebra with algebra operations given by pointwise operations on functions, counit c given by $c(f) = f(e)$, comultiplication

d given by $d(f) = \sum h_i \otimes k_i$ if $f(\alpha\beta) = \sum f_i(\alpha)g_i(\beta)$ for all α, β in $\mathrm{Aut}_\otimes(\mathrm{Mod}(G))$, and antipode $s(f)(\alpha) = f(\alpha^{-1})$ for $\alpha \in \mathrm{Aut}_\otimes(\mathrm{Mod}(G))$.

It is important to unwind what (2.42) actually says: we are given $\mathrm{Mod}(G)$ as a category of vector spaces with tensor product. From that category we produce the group $\mathrm{Aut}_\otimes(\mathrm{Mod}(G))$. From this group and the category we get the set $R(\mathrm{Mod}(G))$ and its Hopf algebra structure. But the Hopf algebra $R(\mathrm{Mod}(G))$ is isomorphic to the Hopf algebra $R(G)$. Thus, the Hopf algebra $R(G)$ is determined from $\mathrm{Mod}(G)$ as a category of vector spaces with tensor product.

We know from (2.29) that G-modules and finite-dimensional $R(G)$-comodules are equivalent categories. This means that objects in $\mathrm{Mod}(G)$ and finite-dimensional $R(\mathrm{Mod}(G))$-comodules are the same. So $\mathrm{Mod}(G)$ is recoverable from $R(\mathrm{Mod}(G))$ and the Hopf algebra $R(\mathrm{Mod}(G))$ is a complete invariant for the category $\mathrm{Mod}(G)$.

These considerations imply that the study of the category $\mathrm{Mod}(G)$ is equivalent to the study of the Hopf algebra $R(\mathrm{Mod}(G))$ and, hence, to the study of the isomorphic Hopf algebra $R(G)$. This seems to be just a restatement of (2.29), where we proved that the category of G-modules could be obtained from the Hopf algebra $R(G)$. But there is a difference, which for our point of view deserves emphasis. The construction of $R(G)$ depends on the group G, while that of $R(\mathrm{Mod}(G))$ depends only on the category $\mathrm{Mod}(G)$. In Example E of Chapter 1, we saw that nonisomorphic groups could have equivalent module categories. This means they have isomorphic groups of tensor automorphisms and, hence, by (2.42) isomorphic Hopf algebras of representative functions. Thus, we must be careful: $\mathrm{Mod}(G)$ determines $R(G)$, but not necessarily G. As we proceed with the study of $R(G)$ (in order to study $\mathrm{Mod}(G)$), we must use the Hopf algebra structure of $R(G)$, rather than view $R(G)$ as functions on G.

This last requirement can be inefficient and counter-intuitive. We would prefer to use modules instead of comodules, and also be able to view elements of $R(G)$ as functions. A convenient way out of these difficulties is to use the group $\mathscr{G}(G)$. First, we observe that $R(G)$ is naturally an algebra of functions on $\mathscr{G}(G)$.

Notation 2.43 Let G be an analytic group and let $f \in R(G)$. Then f defines a function $f: \mathscr{G}(G) \to \mathbb{C}$ by $f(g) = g(f)$.

There are several useful consequences of this observation: if $f \in R(G)$, $f: \mathscr{G}(G) \to \mathbb{C}$ is determined by the restriction of f to $\tau(G)$, since $f(\tau(x)) = f(x)$ for $x \in G$. Further, using the isomorphism $ha: \text{Aut}_\otimes(R(G)) \to \mathscr{G}(G)$ of (2.36) and (2.37), the isomorphism $\sigma^*: R(\text{Mod}(G)) \to R(G)$, and (2.42), we have that $R(G)$ is a Hopf algebra with algebra operations given by pointwise operations of functions on $\mathscr{G}(G)$, counit ε given by $\varepsilon(f) = f(\varepsilon)$, comultiplication Δ given by $\Delta(f) = \sum p_i \otimes q_i$ if $f(gh) = \sum p_i(g)q_i(h)$ for all $g, h \in \mathscr{G}(G)$, and antipode $S(f)(g) = f(g^{-1})$ for $g \in \mathscr{G}(G)$.

Next, we observe that G-modules have $\mathscr{G}(G)$-actions: if V is a G-module, then V has an $\text{Aut}_\otimes(\text{Mod}(G))$-action (2.38), so V has a $\mathscr{G}(G)$-action given by $g(v) = (ha)^{-1}(g)_V(v)$ for $v \in V$ and $g \in G$. We saw in the observations preceding (2.38) that $(ha)\sigma = \tau$, so if $x \in G$ and $v \in V$, $\tau(x)(v) = \sigma(x)v = x(v)$, where the latter is the original G-action on V.

It is elementary to see that a G-action on a finite-dimensional vector space W makes W a G-module if and only if for every $w \in W$ and $\phi \in W^*$, $x \mapsto \phi(x(w))$ is an analytic function on G. So if we begin with a finite-dimensional vector space V with $\mathscr{G}(G)$-action and give it a G-action via τ, then V is a G-module if and only if for each v in V and ϕ in V^*, $x \mapsto \phi(\tau(x)v)$ is in $R(G)$. This suggests the following definition.

Definition 2.44 Let G be an analytic group and let V be a finite-dimensional vector space with $\mathscr{G}(G)$-action. Then V is a $\mathscr{G}(G)$-*module* if for each v in V and ϕ in V^*, $\mathscr{G}(G) \to \mathbb{C}$ by $g \mapsto \phi(g(v))$ is in $R(G)$. A *morphism* of $\mathscr{G}(G)$-modules is a $\mathscr{G}(G)$-equivariant linear transformation. $\text{Mod}(\mathscr{G}(G))$ is the category of $\mathscr{G}(G)$-modules and morphisms.

It is a consequence of our first observations that $\text{Mod}(\mathscr{G}(G))$ is the same as $\text{Mod}(G)$.

Theorem 2.45 Let G be an analytic group. The functor $\tau^*: \text{Mod}(\mathcal{G}(G)) \to \text{Mod}(G)$, which is the identity on morphisms and on objects V, $\tau^*(V) = V$ with G-action via $\tau: G \to \mathcal{G}(G)$ is an isomorphism.

Proof. We do indeed have a functor and it is surjective on objects and morphisms. We need only show that if V is a $\mathcal{G}(G)$-module, $\tau^*(V)$ determines V. Choose a basis $v_1, \ldots, v_n$ of V and dual basis $\phi_1, \ldots, \phi_n$ of V^*. Let $f_{ij}: \mathcal{G}(G) \to \mathbb{C}$ be $f_{ij}(g) = \phi_i(g(v_j))$. By (2.44), $f_{ij} \in R(G)$, and the f_{ij} determine the $\mathcal{G}(G)$-action on V. Since f_{ij} is determined by its restriction to $\tau(G)$, $\tau^*(V)$ determines V.

We are left with the following set up: from the category $\text{Mod}(G)$, we obtain the Hopf algebra $R(G)$ (without its interpretation as functions on G), then the group $\mathcal{G}(G)$ (as the set of $\mathbb{C}$-algebra homomorphisms $R(G) \to \mathbb{C}$ with operations deduced from counit, comultiplication, and antipode of $R(G)$), then we view $R(G)$ as functions on $\mathcal{G}(G)$ and obtain the category of $\mathcal{G}(G)$-modules, which is isomorphic to $\text{Mod}(G)$. Of course, mathematically $R(G)$ remains fundamental, but didactically it will be the study of $\mathcal{G}(G)$ which is most convenient. This study will be undertaken in Chapter 3.

Summary of results of Chapter 2

A representative function on an analytic group G is an analytic function whose translates by G span a finite-dimensional vector space. Every G-module is isomorphic to a submodule of a direct sum of G-modules of representative functions. The set $R(G)$ of all representative functions is an algebra, and a set of algebra generators for $R(G)$ gives a set of G-modules such that every G-module is a submodule of a G-module obtained by tensor products and direct sums from the G-modules in the given set. $R(G)$ is also a Hopf algebra, with counit given by evaluation at e, antipode given by composition with inversion, and comultiplication obtained by the group multiplication of G. Then G-modules are the same as $R(G)$-comodules. Also, representations of G correspond to Hopf algebra homomorphisms to $R(G)$.

Associated to the Hopf algebra $R(G)$ are the groups $\mathscr{G}(G)$ of $\mathbb{C}$-algebra homomorphisms from $R(G)$ to $\mathbb{C}$ and Propaut $R(G)$ of proper automorphisms of $R(G)$. These groups are isomorphic to each other and to the group $\mathrm{Aut}_{\otimes}(\mathrm{Mod}(G))$ of tensor automorphisms of $\mathrm{Mod}(G)$. From the category $\mathrm{Mod}(G)$ and the group $\mathrm{Aut}_{\otimes}(\mathrm{Mod}(G))$ we produce a Hopf algebra of functions on $\mathrm{Aut}_{\otimes}(\mathrm{Mod}(G))$ isomorphic to $R(G)$. Finally, $R(G)$ can be regarded as an algebra of functions on $\mathscr{G}(G)$ and this permits a definition of $\mathscr{G}(G)$-module such that $\mathscr{G}(G)$-modules and G-modules are equivalent.

3
Analytic subgroups of algebraic groups

Let G be an analytic group and $\rho: G \to \mathrm{GL}(V)$ a representation of G. We can regard the Zariski-closure $\bar{G}$ of the image $\rho(G)$ of G as an analytic as well as algebraic group, and then $\rho(G)$ becomes a Zariski-dense analytic subgroup of $\bar{G}$. Such groups have some special properties, which we study in this chapter. We will show that every analytic subgroup of an algebraic group has a normal simply-connected solvable subgroup such that the quotient is reductive and, in fact, this subgroup admits a semidirect complement. If we start with an analytic group with such a semidirect product decomposition, we will show how to embed it as a Zariski-dense normal analytic subgroup of an algebraic group such that the quotient is a torus and, in fact, admits a torus semidirect complement. The existence of this one special embedding can be used to show that the Zariski-closure of the image of any representation is generated by the image of the representation and a torus.

We will be dealing throughout the chapter with Zariski-dense analytic subgroups of algebraic groups. In our first proposition, we record some of the useful basic consequences of this situation.

Proposition 3.1 Let H be an algebraic group and let G be a Zariski-dense analytic subgroup of H. Then:
  (a) If K is a normal analytic subgroup of G, K is a normal analytic subgroup of H;
  (b) $(G, G) = (H, H)$, so in particular G is normal in H and (G, G) is closed in G;
  (c) If H is unipotent, $G = H$.

Proof. (a) Suppose $\rho: H \to \mathrm{GL}(V)$ is an algebraic representation of H, and $W \subseteq U$ are subspaces of V such that $\rho(x)(U) \subseteq W$ for all x in G. Now $T = \{y \in H \mid \rho(y)(U) \subseteq W\}$ is an algebraic subgroup of H containing the Zariski-dense subgroup G, so $T = H$, and $\rho(y)(U) \subseteq W$ for all $y \in H$. If we apply this to the case $V = L(H)$, $\rho = \mathrm{Ad}$, and $U = W = L(K)$, we get that $L(K)$ is $\mathrm{Ad}(H)$-stable and, hence, conclude (a).

(b) We continue to consider the adjoint representation of H. Using $U = L(G)$ and $W = [L(G), L(G)]$, we have $\mathrm{Ad}(G)(U) \subseteq W$, so $\mathrm{Ad}(H)(U) \subseteq W$ or $[L(H), L(G)] \subseteq [L(G), L(G)]$. Using $W = [L(G), L(G)]$ and $U = L(H)$, we have $\mathrm{Ad}(G)(U) \subseteq W$ (since $[L(G), L(H)] \subseteq [L(G), L(G)]$) so $\mathrm{Ad}(H)(U) \subseteq W$ and $[L(H), L(H)] \subseteq [L(G), L(G)]$. This implies (b).

(c) Since $(G, G) = (H, H)$, it will suffice to prove that the canonical injection $G^{ab} \to H^{ab}$ is an isomorphism, so we may assume G and H are abelian. Then H is a vector group and G is a vector subgroup so since G is Zariski-dense in H, $G = H$.

Analytic subgroups of algebraic groups are groups with faithful representations. We are to explore the structure of such groups. The two basic components are introduced in the next two definitions.

Definition 3.2 An analytic group G is *reductive* if every G-module is a direct sum of simple submodules and G has a faithful representation.

Reductive groups, according to our definition, are actually isomorphic to reductive algebraic groups, as we shall now show (this is almost a corollary of (3.1)); we will see later that the isomorphism is even canonical. So the usual structure theory for reductive algebraic groups is available, and will often be used without specific reference.

Proposition 3.3 Let H be an algebraic group and let G be a Zariski-dense analytic subgroup of H. If G^{ab} is a torus, then $G = H$.

Proof. By (3.1)(b), $G^{ab} \to H^{ab}$ is an injection with Zariski-dense image. Now $H^{ab} = V \times T$ where V is a vector group and T is a torus, and the projections of G^{ab} on V and T have Zariski-dense images. Since G^{ab} is a torus, its image in V is 0, so $V = 0$ and $H^{ab} = T$ is a torus. Then $G^{ab} \to H^{ab}_r$ is a Zariski-dense injection of tori, so an isomorphism. Thus, $G/(G, G) = G^{ab} = H^{ab} = H/(G, G)$, so $G = H$.

Proposition 3.4 Let G be a reductive analytic group, $\rho \colon G \to \mathrm{GL}(V)$ a faithful representation and H the Zariski-closure of $\rho(G)$ in $\mathrm{GL}(V)$. Then H is a reductive algebraic group isomorphic analytically to G.

Proof. Let H_0 be the connected component of e in H. Then $G \to H \to H/H_0$ has Zariski-dense image; since G is connected and H/H_0 is discrete, this implies $H = H_0$, so H is connected. By (3.3), $H = \rho(G)$ and, hence, $\rho \colon G \to H$ is an analytic isomorphism. Every algebraic H-module is a G-module via ρ, hence, completely reducible, so H is reductive.

   Proposition 3.4 allows us to make a useful remark about analytic representations of a reductive algebraic group: suppose H is such a group and $\rho \colon H \to \mathrm{GL}(V)$ is an analytic representation. Then $H \times \mathrm{GL}(V)$ is an algebraic group and $\sigma \colon H \to H \times \mathrm{GL}(V)$ given by $\sigma(x) = (x, \rho(x))$ is an analytic homomorphism which is injective, and (3.4) implies that the image $\sigma(H)$ is Zariski-closed in $H \times \mathrm{GL}(V)$ and, hence, the projection of $\sigma(H)$ to $\mathrm{GL}(V)$, namely $\rho(H)$, is Zariski-closed in $\mathrm{GL}(V)$. Now the projection of $\sigma(H)$ to H is an isomorphism of algebraic groups, and its inverse followed by projection to $\mathrm{GL}(V)$ is a morphism of algebraic groups. Thus, ρ is actually an algebraic representation.

   As usual, by a semisimple analytic group we mean a group with semisimple Lie algebra. So every semisimple group satisfies the first condition of (3.2) from standard Lie theory. In fact, it also satisfies the second condition. We will need to use this result in this chapter several times, although the proof would take us far afield at this point. So we state the result as a theorem without proof.

Theorem 3.5 A semisimple analytic group is reductive.

Definition 3.6 Let G be an analytic group. A closed analytic subgroup H of G is a *nucleus* if H is normal, solvable and simply-connected, and G/H is reductive.

Some groups obviously have nuclei: $\mathbb{C}$ is a nucleus of $\mathbb{C}$ and $\{e\}$ is a nucleus of $GL_1\mathbb{C}$. We will ultimately show that an analytic group has a nucleus if and only if it has a faithful representation. The first step will be to show that if G has a nucleus K, then K is actually a semidirect factor of G. This will be proven by considering increasingly general cases, beginning with a small dimensional special case.

Theorem 3.7 Let G be an analytic group and let K be a nucleus of G. Then there is a closed reductive analytic subgroup H of G such that $G = KH$ and $H \cap K = \{e\}$.

Lemma 3.8 Theorem 3.7 is true when $K = \mathbb{C}$ and $G/K = \mathbb{C}^*$.

Proof. Suppose first that G is commutative. We see that G has universal cover $\mathbb{C}^{(2)}$ and fundamental group Z, so $G = \mathbb{C} \times \mathbb{C}^*$. Then there is $\alpha \in \mathbb{C}$ such that $K = \{(x, e^{\alpha x}) \,|\, x \in \mathbb{C}\}$ (K must project onto the first factor; otherwise it would project to 0 on the first factor and inject into the second, which is impossible). Then $H = 0 \times \mathbb{C}^*$ is such that $KH = G$ and $K \cap H = \{e\}$, and H is closed and reductive.

Now assume G is not commutative. Then $L(G)$ is the nonabelian two-dimensional algebra, so the universal cover $\tilde{G}$ of G is the semidirect product of $\mathbb{C}$ with $\mathbb{C}$ acting via the exponential: $\tilde{G} = \mathbb{C} \times_{\exp} \mathbb{C}$. The center of $\tilde{G}$ is $\{0\} \times 2\pi i Z$. We have the covering homomorphism $p: \tilde{G} \to G$ whose kernel is a subgroup of this center isomorphic to $\pi_1(G)$; since G is not simply-connected, this kernel is $\{0\} \times 2\pi i m Z = \Gamma (m \in Z, m \neq 0)$. Now $G = \tilde{G}/\Gamma$ contains the subgroup $H = (0 \times \mathbb{C})/\Gamma$, and H is isomorphic to $\mathbb{C}^*$. The kernel of $H \to G/K$ is $H \cap K$. But since H and G/K are both isomorphic to $\mathbb{C}^*$, this kernel is either finite or all of H. But $H \cap K$ cannot be nontrivial finite (there are no nontrivial finite subgroups of $\mathbb{C} = K$) nor can it be H ($\mathbb{C}^* = H$ is not a subgroup of $\mathbb{C} = K$). Thus, $H \cap K = \{e\}$ and $H \to G/K$ then must be into. So $G = KH$ and $K \cap H = \{e\}$; H is closed since $p^{-1}(H) = 0 \times \mathbb{C}$ is closed in $\tilde{G}$, and reductive.

Lemma 3.9   Theorem 3.7 is true when G is solvable, $K = \mathbb{C}$, and K is central in G.

Proof. Let $n = \dim(G/K)$. Since G/K is solvable and reductive, G/K is a torus of dimension n. We proceed by induction on n. The case $n = 1$ is done in (3.8). So assume the result for $n-1$. Now G/K can be written as $T' \times T''$, where T' is a torus of dimension $n-1$ and $T'' = \mathbb{C}^*$. Let G', G'' be the inverse images of T', T'' in G. Then K is a nucleus of G', and $G'/K = T'$ has dimension $n-1$, so by induction there is a reductive subgroup H' of G' with $G' = KH'$ and $K \cap H' = \{e\}$; H' is a torus of dimension $n-1$. Since K is central in G, $G' = KH'$ is actually a direct product decomposition. Then H' is a characteristic subgroup of G': $H' = \{x \in G' \mid f(x) = 0$ for all $f \in X^+(G')\}$. In a similar fashion we have $G'' = KH''$ with $K \cap H'' = \{e\}$, where H'' is a torus of dimension one, by (3.8). Since G/K is abelian, $(G, G) \subseteq K$ so G' is a normal subgroup of G. Since H' is characteristic in G', H' is also normal in G. In particular, we have a subgroup $H = H'H''$; H is analytic of dimension n, and $H \to G/K$ is surjective (H' maps onto T' and H'' maps onto T''). H is, in fact, a torus: H'' must centralize H' so we have a surjection $H' \times H'' \to H$ and then we have the canonical map $H \to G/K = T' \times T''$ and the composite $H' \times H'' \to T' \times T''$ is an isomorphism, so H is a torus and $H \to G/K$ is an isomorphism. Thus, $G = KH$ and $K \cap H = \{e\}$. This establishes (3.9) for n.

Lemma 3.10   Theorem 3.7 is true when G is solvable and $K = \mathbb{C}$.

Proof. We are going to reduce to the situation of (3.9): We have a homomorphism $S: G \to \text{Aut}(K) = \mathbb{C}^*$ given by $s(g)(x) = gxg^{-1}$. Let G_1 be its kernel and let G_c be the connected component of the identity in G_1. Of course, $K \subseteq G_c$, so we have a surjection $G/K \to G/G_c$. Since G/K is solvable and reductive, it is a torus. So G/G_c is the image of a torus and, hence, is a direct product of a torus and a compact analytic group A. Since G/G_1 is isomorphic to a subgroup of $\text{Aut}(K) = \mathbb{C}^*$, A is in the kernel of $G/G_c \to G/G_1$, which is discrete. So A is trivial and G/G_c is a torus (of dimension one or zero). It follows that the kernel of the surjection $G/K \to G/G_c$, namely G_c/K, is also a torus. Now K is central in G_1, hence G_c, by construction and K is a nucleus of G_c by what we have just shown. So by (3.9), there is a reductive subgroup T of G_c with $G_c = KT$ and $K \cap T = \{e\}$, where T is isomorphic to G_c/K (and therefore a torus). Since K is central in G_c, $G_c = KT$ is actually a direct product and, as in (3.9), we have that T is characteristic in G_c. Since G/K is abelian, $(G, G) \subseteq G_c$ and G_c is normal in G, so T is normal in G. If $G = G_c$ we are done, so assume $G \neq G_c$. Then G/G_c is a one-dimensional torus, which means G_c/K is of codimension one in G/K. We can choose a torus $T' = \mathbb{C}^*$ in G/K such that $G/K = (G_c/K) \times T'$. Let G' be the inverse image of T' in G. By (3.8) applied to G' and its nucleus K, we have a one-dimensional torus H' in G' such that $H' \to G/K$ has T' as its image. Then, $H = TH'$ is a subgroup of G (since T is normal in G) mapping onto G/K. As in (3.8), we see that $T \times H' \to H \to (G_c/K) \times T'$ is an isomorphism, so that H is a torus, $H \cap K = \{e\}$, and $G = KH$, so (3.10) is established.

Lemma 3.11 Theorem 3.7 is true when G is solvable.

Proof. G/K is solvable, hence, a torus. We want to find a torus H in G such that $H \to G/K$ is an isomorphism. Now K is also solvable; let K_0 be the last nonvanishing term in its derived series. Then, K/K_0 is a nucleus of G/K_0. Working by induction on the derived length of K, we can assume that there is a torus $\bar{H}$ in G/K_0 with $\bar{H} \to (G/K_0)/(K/K_0)$ an isomorphism. Let G_0 be the inverse image of $\bar{H}$ in G. Then, K_0 is a nucleus of G_0. If (3.7) holds for G_0, there is a torus H in G_0 with $H \to G_0/K_0$ an isomorphism and, hence, $H \to G/K$ is an isomorphism. We note that K_0 is abelian, so we need only know that (3.7) holds when K is abelian.

If K is abelian, it is a vector group. G acts on K by conjugation, and the action factors through G/K. Since G/K is a torus, acting linearly on the vector group K, there is a one-dimensional subgroup K_0 of K stable under conjugation in G. Thus there is a normal one-dimensional subgroup K_0 of K. By induction on $\dim(K)$ using arguments like the first paragraph of the proof, we reduce to the case $K = K_0$. This is exactly (3.10).

We can now complete the proof of Theorem 3.7. Let R be the radical of G. Then $K \subseteq R$ and R/K is the radical of G/K. Since G/K is reductive, R/K is a torus and, hence, reductive. So K is a nucleus of R. By (3.11), $R = KT$ where $T \cap K = \{e\}$; T is a torus which maps isomorphically to the radical of G/K under $T \to G/K$. Let $L = L(G)$ and $\bar{L} = L(G/K)$; T acts on L via the adjoint representation and stabilizes the kernel $L(K)$ of the surjection $L \to \bar{L}$. We can think of $L \to \bar{L}$ as a morphism of T-modules, and since T is reductive the induced map $L^T \to \bar{L}^T$ is also surjective ($L^T = \{x \in L \mid \mathrm{Ad}(t)(x) = x$ for all $t \in T\}$). Since T maps onto the radical of G/K which is central in G/K, so $\bar{L}^T = \bar{L}$. Now $\bar{L} = Z(\bar{L}) \oplus [\bar{L}, \bar{L}]$ and $[\bar{L}, \bar{L}]$ is semisimple. So we have a Lie algebra surjection $L^T \to \bar{L} \to [\bar{L}, \bar{L}]$ from L^T to a semisimple Lie algebra. This means that there is a semisimple Lie subalgebra L_S of L^T such that $L_S \to [\bar{L}, \bar{L}]$ is an isomorphism. Let S be the analytic subgroup of G with $L(S) = L_S$. Under the projection $G \to G/K$, S maps onto $(G/K, G/K)$ inducing an isomorphism on Lie algebras. By (3.5) and (3.4), S has finite center, and this finite group contains the kernel of $S \to (G/K, G/K)$, which is $S \cap K$. Since K is simply-connected solvable, it has no finite subgroups, so $S \cap K = \{e\}$. This means that $S \to G/K$ is an injection with image $(G/K, G/K)$. Now $L(S) \subseteq L^T$ which is the Lie algebra of $G^T = \{x \in G \mid xt = tx$ for all t in $T\}$, so S and T commute. Let H be the group TS; H is a subgroup of G, and since $G/K = R/K(G/K, G/K)$, H maps onto G/K. Under $S \to G/K$, $S \cap R$ maps to $R/K \cap (G/K, G/K)$, which is finite. We want to show $H \cap K = \{e\}$. Let $x = ts$ be in $H \cap K$, where $t \in T$ and $s \in S$. Since $T \subseteq R$ and $K \subseteq R$, $s \in S \cap R$, so there is a positive integer n with $s^n = e$. Now s and t commute, so $x^n = s^n t^n = t^n$ is in $K \cap T = \{e\}$. Thus, x is an element of K of finite order, hence, trivial. We have shown $H \cap K = \{e\}$, so $H \to G/K$ is an isomorphism, and $G = KH$ with $K \cap H = \{e\}$. We have produced the desired reductive subgroup of G, and the proof of (3.7) is complete.

Theorem 3.7 was originally proved by Hochschild & Mostow by using the fact that a reductive analytic group is the complexification of a compact real Lie group. They then could lift the compact subgroup of G/K to a compact subgroup of G, and its complexification is the desired group H. Of course, the fact alluded to is hardly an immediate consequence of Definition 3.2 of reductive group.

Now suppose G is an analytic group with nucleus K. We want to construct a special type of faithful representation of G, or equivalently, embed G as an analytic subgroup of an algebraic group. Using (3.7), we can write $G = KH$, where H is reductive and $K \cap H = \{e\}$. We can try to construct the embedding by doing K and H separately, and this will be our program. We begin by introducing some terminology for our goal:

Definition 3.12 Let G be an analytic group. A *split hull* of G is a triple $(\bar{G}, f, T)$ where $\bar{G}$ is an algebraic group, $f: G \to \bar{G}$ is an analytic injection with Zariski-dense image, and T is a torus in $\bar{G}$ with $\bar{G} = f(G)T$ and $G \cap T = \{e\}$. (By (3.1(b)), $f(G)$ is normal in $\bar{G}$.)

We will ultimately show in (3.16) that every analytic group with a nucleus has a split hull. The existence of a split hull is the key to much of the subsequent development. The sudden appearance of the torus in the split hull may be a bit mysterious, so before proceeding to the proof of (3.16) we will discuss where the torus comes from.

As mentioned above, we will construct a split hull of the analytic group G with nucleus K by constructing one for K, and a reductive semidirect complement to K, separately. The torus appears in the embedding for K; K is simply-connected and solvable, but not necessarily nilpotent. If $\bar{K}$ is the Zariski-closure of K in a faithful representation, then $\bar{K}$ is solvable but need not be nilpotent, so $\bar{K}$ is a solvable nonunipotent algebraic group and, hence, contains a torus. So tori appear. We can be a bit more precise: if K is not nilpotent, it has a proper Cartan subgroup C. Then $K = (K, K)C$; both (K, K) and C are simply-connected nilpotent (so can be regarded as unipotent algebraic groups), and C acts on the normal subgroup (K, K) by conjugation. This gives a homomorphsim $s: C \to \operatorname{Aut}((K, K)) = \operatorname{Aut}(L((K, K))) \subseteq \operatorname{GL}(L((K, K)))$, the latter containment being of algebraic groups. The Zariski-closure $\overline{s(C)}$ of $s(C)$ in $\operatorname{Aut}((K, K))$ is algebraic and we can use it to form the semidirect product of (K, K) with $\overline{s(C)}$, which is an algebraic group. If this is done correctly, we get an algebraic group containing

K. Now $\overline{s(C)}$ is nilpotent, but not unipotent, and its maximal torus becomes the torus of the split hull.

With these remarks in mind, we start the construction of a split hull. This will be done in a sequence of lemmas treating increasingly general cases.

Lemma 3.13 Let K be a simply-connected solvable analytic group which is the semidirect product of the analytic subgroups N and C, with N normal in K and N and C nilpotent. Suppose that a is an analytic automorphism of K with $a(N) = N$ and $a(c) = c$ for all c in C. Then K has a split hull $(\bar{K}, f, T)$ and there is a unique algebraic automorphism $\bar{a}$ of $\bar{K}$ with $\bar{a}f = fa$; moreover, $\bar{a}(t) = t$ for all t in T.

Proof. We can regard N (respectively, C) as unipotent algebraic groups such that $\exp_N$ (respectively, $\exp_C$) is an isomorphism of the algebraic variety N (respectively, C) with the linear variety $L(N)$ (respectively, $L(C)$). Then $\mathrm{Aut}(N) = \mathrm{Aut}(L(N))$, the group of analytic group automorphisms of N, is an algebraic group. We have an analytic group homomorphism $s: C \to \mathrm{Aut}(N)$ given by $s(c)(n) = cnc^{-1}$. Let $\overline{s(C)}$ be the Zariski-closure of $s(C)$. Now C is nilpotent, hence, so is $\overline{s(C)}$ and, hence, $\overline{s(C)} = U \times S$ where S is a torus and U is unipotent. Let $p: \overline{s(C)} \to U$ and $q: \overline{s(C)} \to S$ be the projections. Let D be the algebraic group $C \times S$ and let $T: D \to \mathrm{Aut}(N)$ be given by $t(c, x) = (ps(c))x$. Clearly t is a homomorphism; it is actually a homomorphism of algebraic groups: for t factors as $C \times S \to U \times S \to \mathrm{Aut}(N)$ where the last map is inclusion and the first sends (c, s) to $(ps(c), x)$. Since $ps: C \to U$ is an analytic homomorphism of unipotent groups, it is algebraic. Let $\bar{K}$ be the algebraic group $N \times_t D$. Define $f: K \to \bar{K}$ by $f(x) = (n, (c, qs(c)))$ if $x = nc$ is the unique representation of x with $n \in N$ and $c \in C$. It is easy to see that f is a group homomorphism, and f is analytic and injective. Let $T \subseteq \bar{K}$ be the torus $\{e\} \times (\{e\} \times S)$. We have $f(K)T = \bar{K}$ and $f(K) \cap T = \{e\}$. Now $f(N) = N \times \{e\}$ is Zariski-closed in $\bar{K}$; to show that $f(K)$ is Zariski-dense in $\bar{K}$ it will suffice to show that $f(C)$ is Zariski-dense in $\{e\} \times D$, for then the Zariski-closure of $f(K)$ will contain both $N \times \{e\}$ and $\{e\} \times D$ and, hence equal $\bar{K}$. To show that $f(C)$ is dense in $\{e\} \times D$, we consider $h: C \to D$ given by $h(c) = (c, qs(c))$. Let $\overline{h(C)}$ be the Zariski-closure of $h(C)$ in D. Now C, and hence $\overline{h(C)}$, is nilpotent, so $\overline{h(C)} = V \times R$ where V is the unipotent radical and R the maximal torus of $\overline{h(C)}$. Since the projection $\overline{h(C)} \to C$ (restriction of the projection $D = C \times S \to C$) is onto, V projects onto C. The projection from $h(C)$ to S has $qs(C)$ as its image, which is dense in S, so $\overline{h(C)}$ and, thus R, projects onto S. So $\overline{h(C)} = D$. Since $f(c) = (e, h(c))$ for c in C, $f(C)$ is dense in $\{e\} \times D$.

We have now shown that $(\bar{K}, f, T)$ is a split hull of K. Now we consider the automorphism a. Define $\bar{a}: \bar{K} \to \bar{K}$ by $\bar{a}(n, d) = (a(n), d)$ for $n \in N$, $d \in D$ (this makes sense since $a(N) = N$). Then $\bar{a}$ is an automorphism of $\bar{K}$ as algebraic variety, with $\bar{a}f = fa$, and uniquely determined by this condition since $f(K)$ is dense in $\bar{K}$. We need to see that $\bar{a}$ is a group automorphism. A calculation in the semidirect product shows that $\bar{a}$ is a group automorphism if $a(t(d)n) = t(d)a(n)$ for all d in D and n in N; i.e., that $a|N$ commutes with $t(D)$. Using the map $h: C \to D$ of the paragraph above, we have that for $c \in C$, $th(c) = t(c, qs(c)) = ps(c)qs(c) = s(c)$. Now $s(c)$ and $a(N)$ commute since a is the identity on C. Thus, the Zariski-dense subgroup $t(h(C))$ of D commutes with $a|N$, so $t(D)$ does also. By construction, $\bar{a}$ is the identity on $\{e\} \times D$ and, hence, also on T.

The requirement that K be a semidirect product in (3.13) can be weakened, which we now show.

Lemma 3.14 Let K be a simply-connected solvable analytic group and let N and C be nilpotent analytic subgroups of K with N normal such that $K = NC$. Suppose a is an analytic automorphism of K with $a(N) = N$ and $a(c) = c$ for all c in C. Then K has a split hull $(\bar{K}, f, T)$ and there is a unique algebraic automorphism $\bar{a}$ of $\bar{K}$ with $\bar{a}f = fa$; moreover, $\bar{a}(t) = t$ for all t in T.

Proof. As in (3.13), N and C are unipotent algebraic groups and we have an analytic homomorphism $s: C \to \text{Aut}(N)$ where $s(c) \times (n) = cnc^{-1}$. We form the analytic group $K' = N \times_s C$. We have an analytic automorphism a' of K' where $a'(n, c) = (a(n), c)$. Then

(3.13) applies to K' to yield a split hull $(\bar{K}', f', T')$ and an automorphism $\bar{a}'$ of $\bar{K}'$.

There is an analytic epimorphism $g: K' \to K$ given by $g(n, c) = nc$. Let $L = \mathrm{Ker}(g)$ and let L_0 be the connected component of e in L. Then $K'/L_0 \to K$ has the discrete group L/L_0 as kernel; since K is simply-connected $L/L_0 = \{e\}$ and L is connected. In fact, $L = \{(x, x^{-1}) \mid x \in N \cap C\}$ so L is also nilpotent.

We want to examine $f'(L)$. First, consider $s(N \cap C)$. Under the identification $\mathrm{Aut}(N) = \mathrm{Aut}(L(N))$, $s(x) = \mathrm{Ad}(x)$ if $x \in C$. Since N is nilpotent, if $x \in N \cap C$, then $\mathrm{Ad}(x)$ is unipotent. This shows that $s(N \cap C)$ is contained in the unipotent radical of the Zariski-closure of $s(C)$. In terms of the map $h: C \to D$ of the proof of (3.13) (here, of course, $C = \{e\} \times C \subseteq K'$) this means that $h(N \cap C) = (N \cap C) \times \{e\}$. If $y = (x, x^{-1}) \in L$ with $x \in N \cap C$, then $f'(y) = (x, (x^{-1}, e))$. This means that $f'(L)$ is in the unipotent radical of $\bar{K}'$, so by (3.1)(c) $f'(L)$ is Zariski-closed in $\bar{K}'$; L is normal in K' so by (3.1)(a) $f'(L)$ is normal in $\bar{K}'$. We can thus form the algebraic group $\bar{K} = \bar{K}'/f'(L)$, and we get an induced analytic map $f: K = K'/L \to \bar{K}$. Now f is injective and $f(K)$ is Zariski-dense in $\bar{K}$ since $f'(K')$ is Zariski-dense in $\bar{K}'$. If T is the image of T' in $\bar{K}$, T is a torus with $f(K)T = \bar{K}$ and $f(K) \cap T = \{e\}$ (the latter since $T' \cap f'(L) = \{e\}$). So $(\bar{K}, f, T)$ is a split hull of K. Moreover, $\bar{a}'$ is the identity on $f'(L)$ and, hence, induces an automorphism $\bar{a}$ of $\bar{K}$ which has the desired properties.

If G is an analytic group that has a nucleus K, and H is a reductive subgroup of G with $G = KH$ and $K \cap H = \{e\}$, then every element of H induces, by conjugation, an automorphism of K. We want to show that K has subgroups N and C as in (3.14) such that the automorphisms of K induced from H satisfy the conditions of (3.14). Then we will use (3.14) to produce a split hull of G. The existence of N and C comes from the following proposition:

Proposition 3.15 Let G be an analytic group, K a nucleus of G, and H a reductive subgroup of G with $G = KH$ and $K \cap H = \{e\}$. Then there are nilpotent analytic subgroups N and C of K with $K = NC$ such that N is normal in G and $(C, H) = \{e\}$.

Proof. H acts on $L(K)$ via the adjoint representation, and since H is reductive the H-module $L(K)$ is a direct sum of $L(K)^H$ and a submodule S such that $\text{Ad}(H)(S) = S$. Hence, $S \subseteq [L(H), S] \subseteq [L(G), L(K)]$. Let N be the analytic subgroup of K with $L(N) = [L(G), L(K)]$. Then N is a nilpotent subgroup of K normal in G. Let C be an analytic subgroup of K such that $L(C)$ is a Cartan subalgebra of $L(K)^H$; C is nilpotent and $(C, H) = \{e\}$. Since $L(C)$ is Cartan in $L(K)^H$, $L(K)^H = L(C) + [L(K)^H, L(K)^H] \subseteq L(C) + L(N)$. Also $L(K) = L(K)^H + S = L(C) + L(N)$, so $K = NC$.

We are now in a position to establish that every analytic group with a nucleus has a split hull.

Theorem 3.16 Let G be an analytic group with a nucleus K, and let H be a reductive analytic subgroup of G with $G = KH$ and $K \cap H = \{e\}$. Then G has a split hull $(\bar{G}, f, T)$ with $(T, f(H)) = \{e\}$.

Proof. Let N and C be nilpotent analytic subgroups of K with N normal in G, $K = NC$, and $(H, C) = \{e\}$ (N and C exist by (3.15)). For $h \in H$, let $a(h)$ be the automorphism of K given by $a(h)(k) = hkh^{-1}$. Then $a(h)(N) = N$ and $a(h)$ is the identity on C. By (3.14), K has a split hull $(\bar{K}, g, S)$ and there is a unique algebraic automorphism $\bar{a}(h)$ of $\bar{K}$ with $\bar{a}(h)g = ga(h)$ and $\bar{a}(h)$ is the identity on S. It follows that $\bar{a}: H \to \text{Aut}(\bar{K})$ (algebraic group automorphisms) is a homomorphism. Let $\bar{G}$ be the semidirect product $\bar{K} \times_{\bar{a}} H$. Let $f: G \to \bar{G}$ be given by $f(x) = (g(k), h)$ if $x = kh$ is the unique representation of x with $k \in K$ and $h \in H$. Let $T = f(S)$. We have that $\bar{G}$ is an analytic group, f is an analytic embedding, T is a torus, $\bar{G} = GS$ and $G \cap S = \{e\}$, and $(T, f(H)) = \{e\}$ from the properties of $(\bar{K}, g, S)$. We still must see that $\bar{G}$ is an algebraic group and G is Zariski-dense in $\bar{G}$; $\bar{G}$ will be an algebraic group if H acts algebraically on $\bar{K}$, i.e., if $H \times \bar{K} \to \bar{K}$ by $(h, x) \mapsto \bar{a}(h)(x)$ is algebraic. To see this, we will need to know that S is a maximal torus of $\bar{K}$. Since $\bar{K}$ is solvable, $(\bar{K}, \bar{K})$ is contained in the unipotent radical U of $\bar{K}$, so it will be enough to show that S maps onto the maximal torus of $\bar{K}^{ab}$. But $\bar{K}^{ab} = K^{ab} \times S$ and K^{ab} is simply-connected, so $S = \text{Ker}(X^+(\bar{K}^{ab}))$, which proves S is maximal. Thus, as algebraic groups, $\bar{K}$ is the semidirect product of U and S, so as varieties

$\bar{K} = U \times S$. Then $H \times \bar{K} \to \bar{K}$ is $H \times U \times S \to U \times S$ by $(h, u, s) \mapsto \bar{a}(h)(us) = (\bar{a}(h)u, s)$, the last equality since $\bar{a}(h)$ is the identity on S. So we are reduced to showing that $\bar{a}$ gives an algebraic action $H \times U \to U$. Since U is simply-connected nilpotent, this is the same as the assertion that $\bar{a} : H \to \operatorname{Aut}(U) = \operatorname{Aut}(L(U))$ is algebraic, and since H is reductive it is enough to know that this map is analytic. It is even sufficient for $\bar{a} : H \to \operatorname{Aut}(L(\bar{K}))$ to be analytic. Now $L(\bar{K}) = L(K) + L(S)$ (semidirect product of Lie algebras with $L(K)$ the ideal), and $\bar{a}(H)$ acts trivially on $L(S)$, so $\bar{a} : H \to \operatorname{Aut}(L(\bar{K}))$ is really just the adjoint representation of H on $L(K)$ extended by zero, and hence analytic. So $\bar{G}$ is an algebraic group.

Now $f(H)$ is Zariski-closed in $\bar{G}(f(H) = \{e\} \times H)$, and $f(K) \subseteq \bar{K} \times \{e\}$ and is Zariski-dense. Thus, the Zariski-closure of $f(G)$ contains $\bar{K} \times \{e\}$ and $\{e\} \times H$, so equals $\bar{G}$. This completes the proof of (3.16).

We note that (3.7) and (3.16) combine to show that an analytic group with a nucleus is an analytic subgroup of an algebraic group and, hence, has a faithful representation. The converse is also true: groups with faithful representations have nuclei, as we will see in (3.23).

Both (3.7) and (3.16) have involved proofs. They are, in fact, the cornerstone of the whole representation theory of analytic groups, as the patient reader will eventually discover. Because we have placed them at the beginning of the theory, the proofs are almost 'first principles' arguments, which accounts somewhat for their length. In a very real sense, the rest of this work can be regarded as an exploration of the consequences of (3.16).

We also want to comment on the degree to which the split hull of (3.16) is canonical. The torus T appears from the subgroup C of K from (3.13). Once K and C are specified, the construction is canonical, N being just (G, K). But the group G can have nuclei which look quite different and yield distinct split hulls. Here is a concrete example:

Example. Let G be the (algebraic) group $\mathbb{C} \times \mathbb{C} \times \mathbb{C}^*$ with operation $(z, x, t)(z', x', t') = (t + z', x + tx', tt')$. Let $K = \mathbb{C} \times \mathbb{C} \times \{1\}$

(the unipotent radical of G) and let $K' = \{(z, x, e^z) \mid x, z \in \mathbb{C}\}$. If $H = \{(0, 0)\} \times \mathbb{C}^*$, $G = KH = K'H$ and $K \cap H = K' \cap H = \{e\}$, and both K and K' are nuclei of G, H being reductive. Then $L(G)$ has basis A, B, C where $\exp_G(zA) = (z, 0, 0)$, $\exp_G(xB) = (0, x, 0)$ and $\exp_G(tC) = (0, 0, e^t)$. Then $L(H) = \langle C \rangle$, $L(K) = \langle A, B \rangle$ and $L(K') = \langle A + C, B \rangle$. The operation in $L(G)$ is $[A, B] = [A, C] = 0$ and $[C, B] = B$. Thus, $[L(G), L(K)] = \langle B \rangle = [L(G), L(K')]$, $\{X \in L(K) \mid [L(H), X] = 0\} = \langle A \rangle$, and $\{X \in L(K') \mid [L(H), X] = 0\} = \langle A + C \rangle$. Thus, $N = (G, K) = (G, K') = \{0\} \times \mathbb{C} \times \{1\}$, $C = K^H = \mathbb{C} \times \{0\} \times \{1\}$ and $C' = (K')^H = \{(z, 0, e^z \mid z \in \mathbb{C}\}$. Thus, to carry out the construction of a split hull using (3.16), with K we write $K = NC$ and with K' we write $K' = NC'$. The maps $s \colon C \to \mathrm{Aut}(N) = \mathbb{C}^*$ and $s' \colon C' \to \mathrm{Aut}(N) = \mathbb{C}^*$ are then $s((z, 0, 1)) = 1$ and $s'((z, 0, e^z)) = e^z$. Thus, the torus arising from K is trivial, while that from K' is not, the torus being the maximal torus in the closure of the image of s, s', respectively. Thus, the split hulls are $(G, \mathrm{id}, \{1\})$ (from K) and $(\bar{G}, f, T)$ (from K'), where $\bar{G} = \mathbb{C} \times \mathbb{C} \times \mathbb{C}^* \times \mathbb{C}^*$ with operation $(z, x, t, s)(z', x', t', s') = (z + z', x + tx', tt', ss')$ and $f \colon G \to \bar{G}$ is $f(z, x, t) = (z, x, t, e^z)$ where $T = \{(0, 0, t, t) \mid t \in \mathbb{C}^*\}$.

We next establish a permanence property of split hulls under inverse images.

Theorem 3.17 Let G be an analytic group, let $(\bar{G}, f, T)$ be a split hull of G and let $h \colon G' \to \bar{G}$ be a homomorphism of algebraic groups. Suppose $f' \colon G \to G'$ is an analytic homomorphism with Zariski-dense image such that $hf' = f$. Then $T' = h^{-1}(T)$ is a torus in G' and (G', f', T') is a split hull of G.

Proof. First, we show that $G' = f'(G)T'$ and $f'(G) \cap T' = \{e\}$. $f'(G)T'$ is a subgroup of G' since $f'(G)$ is normal in G', (3.1)(a), and $h(f'(G)T') = f(G)T = \bar{G}$. Since $\mathrm{Ker}(h) \subseteq T' \subseteq f'(G)T'$, $f'(G)T' = G'$. If $x \in f'(G) \cap T'$, $h(x) \in f(G) \cap T = \{e\}$, so $x \in \mathrm{Ker}(h) \cap f'(G)$. If $x = f'(y)$ with $y \in G$, then $h(x) = hf'(y) = f(y) = e$, and since f is injective this means $y = e$ so $x = e$.

Next, we show that T' is a torus. This will first be done in the case that G is commutative. Then $f(G)$, $f'(G)$ are Zariski-dense commutative subgroups of $\bar{G}$, G', so $\bar{G}$, G' are commutative. Let $\bar{G} = U \times S$, $G' = U' \times S'$, where U, U' are unipotent radicals and

S, S' are maximal tori. Now $f(G) = V \times S_0$ where V is a vector group and S_0 is a torus. As analytic groups, $\bar{G} = f(G) \times T = V \times S_0 \times T$. Therefore $S = S_0 \times T$. Counting dimensions, we see $\dim V = \dim U = \mathrm{rank}(X^+(G)) = m$. Consider $G \to U'$ given by f' followed by projection. This map has Zariski-dense image, so the image is U' by (3.1)(c), so $\dim U' \leq m$. Since $h(U') = U$ and $\dim U' \leq \dim U$, $h: U' \to U$ is an isomorphism. It follows that $h^{-1}(S) = S'$. Thus, $T' \subseteq S'$. Since $G' = f'(G) \times T'$, T' is connected and, hence, a torus.

Now we no longer assume G commutative. We have a commutative diagram

By (3.1)(b), $(\bar{G}, \bar{G}) = (f(G), f(G))$ and $(G', G') = (f'(G), f'(G))$. Since $f'(G) \cap T' = \{e\}$, and $f(G) \cap T = \{e\}$, $g'|T'$ and $g|T$ are injections, and $(G'^{ab}, g'f', g'(T'))$, $(\bar{G}^{ab}, gf, g(T))$ are split hulls of G^{ab}. Now $(g')^{-1}(\bar{h})^{-1}(g(T)) = h^{-1}g^{-1}(g(T)) = h^{-1}(T(\bar{G}, \bar{G})) = T'(G', G')$. Thus, $g'(T') = g'(T'(G', G')) = g'((g')^{-1}(\bar{h})^{-1}(g(T))) = (\bar{h})^{-1}(g(T))$. By the commutative case just considered, $(\bar{h})^{-1}(g(T))$ is a torus. Thus, $g'(T')$ is a torus, so T' is a torus. This proves that (G', f', T') is a split hull of G.

Using (3.17), we can realize every algebraic group to which the analytic group G maps via an analytic homomorphism with Zariski-dense image as an algebraic image of a split hull of G. For later use we are going to state this with a little more generality:

Proposition 3.18 Let G be an analytic group, $(\bar{G}, f, T)$ a split hull of G, and $p: G \to H$ an analytic homomorphism with Zariski-dense image to the algebraic group H. Then there is a split hull (G', f', T') of G and algebraic homomorphisms $g: G' \to \bar{G}$ and $h: G' \to H$ such that $gf' = f$, $hf' = p$, and $T' = g^{-1}(T)$, with h surjective.

Proof. $\bar{G} \times H$ is an algebraic group and $f': G \to \bar{G} \times H$ given by $f'(g) = (f(g), p(g))$ is an injective analytic homomorphism. Let G'

be the Zariski-closure of $f'(G)$, and let $g: G' \to \bar{G}$ and $h: G' \to H$ be the restrictions of the projections of $\bar{G} \times H$ on $\bar{G}$ and H. These are algebraic homomorphisms such that $gf' = f$ and $hf' = p$. Let $T' = g^{-1}(T)$. By (3.17), (G', f', T') is a split hull of G. The image of h is Zariski-closed and contains $p(G)$, so h is surjective.

Corollary 3.19 Let G be an analytic group with a split hull, let H be an algebraic group and let $p: G \to H$ be an analytic homomorphism with Zariski-dense image. Then there is a torus S in H such that $H = p(G)S$ and $L(H) = L(p(G)) \oplus L(S)$.

Proof. By (3.18) there is a split hull (G', f', T') of G and a surjective algebraic homomorphism $h: G' \to H$ such that $hf' = p$. Let $S' = h(T')$. Then S' is a torus and $H = h(G') = h(f'(G)T') = p(G)S'$. So $L(H) = L(p(G)) + L(S')$ (the sum is not necessarily direct). Let $X = \text{Ker}(\exp: L(S') \to S')$ and let B be a basis of X (hence, also of $L(S')$). Now $L(S') \to L(H)/L(p(G))$ is onto, so there is a subset C of B mapping to a basis of $L(H)/L(p(G))$. The C-span of C is the Lie algebra of a subtorus S of S' such that $L(H) = L(p(G)) \oplus L(S)$ and, hence, also $H = p(G)S$.

In the notation of (3.19), if p is injective and $S \cap p(G) = \{e\}$, then (H, p, S) is a split hull of G. Since $L(p(G)) \cap L(S) = \{0\}$, we always have that $S \cap p(G)$ is discrete in $p(G)$. This tells us that every faithful representation is almost a split hull: the only question is if the discrete group $S \cap p(G)$ is trivial.

If G is an algebraic group, G has a split hull, namely $(G, 1, \{e\})$. We can refine (3.19) in this case.

Corollary 3.20 Let G be an algebraic group, let H be an algebraic group and let $p: G \to H$ be an analytic homomorphism with Zariski-dense image. Then there is a torus S in H such that $H = p(G)S$, $L(H) = L(p(G)) \oplus L(S)$, and $(S, p(G)) = \{e\}$.

Proof. By (3.18), there is a split hull (G', f', T') of G, a surjective algebraic homomorphism $h: G' \to H$ such that $hf' = p$, and an algebraic homomorphism $g: G' \to G$ such that $gf' = 1$ and $T' = g^{-1}(e)$ (we are using $(G, 1, \{e\})$ for the split hull $(\bar{G}, f, T)$ of (3.18)). So T'

is normal in G', and hence central. Thus, $(h(T'), p(G)) = \{e\}$. The torus S produced in (3.19) is a subtorus of $h(T')$, so $(S, p(G)) = \{e\}$, and the other assertions were established in (3.19).

When we apply the results of this chapter to rings of representative functions, which we will do in Chapter 4, we will need to know about mapping analytic groups to inverse systems of algebraic groups. Theorem 3.17 provides the necessary information here, when the inverse system is directed and contains a split hull of the analytic group:

Theorem 3.21   Let G be an analytic group and let $\{H_i, f_{ij}: H_j \to H_i \mid i, j \in I\}$ be a directed inverse system of algebraic groups over the directed set I. Suppose that for each $i \in I$ there is an analytic homomorphism $p_i: G \to H_i$ such that $f_{ij}p_j = p_i$, and that there is an $i_0 \in I$ and a torus $T_{i_0} \subseteq H_{i_0}$ such that $(H_{i_0}, p_{i_0}, T_{i_0})$ is a split hull of G. Let H be the inverse limit of the system, let $p: G \to H$ be induced from the p_i, and let $f_i: H \to H_i$ be the projection. Then there is a subgroup T of H such that $H = p(G)T$ and $T \cap p(G) = \{e\}$; $p(G)$ is normal in H, p is injective, and for each $i \in I$, f_i is surjective and $f_i(T)$ is a torus in H_i.

Proof. Because I is directed, we can form the inverse limit over the cofinal subset I' of I of elements dominating i_0, and it will suffice to prove the theorem with I' replacing I. So we can assume i_0 is minimal in I. For each $i \in I$, let $T_i \subseteq H_i$ be $f_{i_0i}^{-1}(T_{i_0})$. By (3.17), (H_i, p_i, T_i) is a split hull of G. If $j \geq i$, $f_{ij}^{-1}(T_i) = T_j$, and $f_{ij}(p_j(G)) \subseteq p_i(G)$ and $f_{ij}(T_j) \subseteq T_i$. Let T be the inverse limit of $\{T_i, f_{ij}: T_j \to T_i \mid i, j \in I\}$. Now H is a subgroup of $\prod\{H_i \mid i \in I\} = H_0$, which contains the subgroups $\prod\{p(G_i) \mid i \in I\} = G_0$ and $\prod\{T_i \mid i \in T\} = T_0$. Since $H_i = p(G_i)T_i$, $p(G_i)$ normal in H_i and $T_i \cap p(G_i) = \{e\}$, G_0 is normal in H_0, $T_0 \cap G_0 = \{e\}$ and $G_0T_0 = H_0$. Elementary arguments with inverse limits then show that $p(G)T = H$, $p(G) \cap T = \{e\}$ and $p(G)$ is normal in H, when T is regarded as a subgroup of H. Now $f_{ij}: p_j(G) \to p_i(G)$ is an isomorphism, so $f_i: p(G) \to p_i(G)$ is an isomorphism for all i. Since $H_i = p_i(G)T_i$, to complete the proof we need only show that $f_i(T) = T_i$. This will be recorded in a separate lemma:

Lemma 3.22 Let $\{T_i, f_{ij}: T_j \to T_i \,|\, i, j \in I\}$ be a directed inverse system of tori and surjective algebraic homomorphisms, let $T = \text{proj lim}\{T_i \,|\, i \in I\}$, and let $f_i: T \to T_i$ be the induced maps. Then each f_i is surjective.

Proof. Let A_i be the affine coordinate ring of T_i and for $j > i$ let $p_{ji}: A_i \to A_j$ be the injective ring homomorphism induced from $f_{ij}: T_j \to T_i$. Let $A = \text{dir lim}\{A_i \,|\, i \in I\}$. Replace each A_i by its image in A, so $A = \bigcup\{A_i \,|\, i \in I\}$. Let $X_i = X(T_i)$; X_i is a multiplicative (abelian) subgroup of A_i and A_i is the complex group ring $\mathbb{C}[X_i]$. If $X = \bigcup X_i$, then A is the complex group ring $\mathbb{C}[X]$. Since T_i is the group of $\mathbb{C}$-algebra homomorphisms $A_i \to \mathbb{C}$, T is the group of $\mathbb{C}$-algebra homomorphisms $A \to \mathbb{C}$. To show that $f_i: T \to T_i$ is surjective, we need to show that every $\mathbb{C}$-algebra homomorphism from A_i to $\mathbb{C}$ can be extended to a $\mathbb{C}$-algebra homomorphism from A to $\mathbb{C}$. Since we are dealing with group rings, this means showing that every group homomorphism from X_i to $\mathbb{C} - \{0\}$ can be extended to a homomorphism from X to $\mathbb{C} - \{0\}$, and this in turn follows from the divisibility of the group $\mathbb{C} - \{0\}$. Thus (3.22) (and, hence, (3.21)) follows.

Let us note for future reference that in (3.22) we identified T with the group of $\mathbb{C}$-algebra homomorphisms from $\text{dir lim}(\mathbb{C}[T_i]) = A$ to $\mathbb{C}$, and that A was the complex group ring $\mathbb{C}[\text{dir lim } X(T_i)]$.

All the major results of this chapter focus on groups with nuclei. It is time to give a representation theoretic characterization of these groups.

Theorem 3.23 Let G be an analytic group with a faithful representation. Then G has a nucleus.

Proof. The hypothesis implies that G can be regarded as a Zuriski-dense analytic subgroup of an algebraic group H (H is the Zariski-closure of $p(G)$ if $p: G \to \text{GL}(V)$ is a faithful representation). Let R be the radical of G.

We first assume R is central in G and, hence, also in H. Then $R = V \times T$, where V is a vector group and T is a torus. Let S be a semisimple subgroup of G such that $G = RS$. Then $R \cap S$ is central in S, and by (3.5) this means $R \cap S$ is finite. If $x \in V \cap TS$, say $x = ts$ with $t \in T$ and $s \in S$, then $s = t^{-1}x \in R \cap S$ (since $x \in V$), so has finite order. Since $R = V \times T$ and V has no elements of finite order, this means $x = e$. Let $P = TS$. By (3.4), S is closed (and reductive) in H, and hence, P is reductive (in H) so P is a reductive group. Now $G = VP$ and $V \cap P = \{e\}$ with V normal in G, so V is actually a nucleus of G.

Now we return to general G. Let $\bar{R}$ be the Zariski-closure of R in H. Then $\bar{R}$ is in the radical of H so $(\bar{G}, \bar{R})$ is in the unipotent radical U of H. Thus, (G, R) is in U. By (3.1)(c), (G, R) is Zariski-closed in H and by (3.1)(a), (G, R) is normal in H. Then $H/(G, R)$ is an algebraic group containing $G/(G, R)$ as a Zariski-dense analytic subgroup. The radical $R/(G, R)$ of $G/(G, R)$ is central, so $G/(G, R)$ has a nucleus V from the case previously considered. The inverse image K of V in G is a closed normal subgroup such that G/K is reductive. Now K contains the normal simply-connected nilpotent subgroup (G, R) and $K/(G, R) = V$ is also simply-connected (and abelian), so K is simply-connected and solvable. Thus, K is a nucleus of G.

As noted previously, after (3.16), (3.7) and (3.16) show that an analytic group with a nucleus has a faithful representation. So having a nucleus is equivalent to having a faithful representation. Having a nucleus implies having a split hull, and having a faithful representation implies being an analytic subgroup of an algebraic group, so every analytic subgroup of an algebraic group has a split hull. Now the image of any analytic group under a homomorphism to an algebraic group is an analytic subgroup of an algebraic group. We can apply this to strengthen (3.19).

Proposition 3.24 Let G be an analytic group, H an algebraic group, and $p: G \to H$ an analytic homomorphsim with Zariski-dense image. Then there is a torus T in H with $H = p(G)T$, and $L(H) = L(p(G)) \oplus L(T)$.

Proof. Apply (3.19) to the analytic group $p(G)$, which has a split hull by the above remarks.

If we analyze the proof of (3.23), we see that G will have a nucleus and, hence, a faithful representation, if (G, R) is simply-connected and $R/(G, R)$ has a faithful representation (R is the radical of G). We use this technique to establish the following criterion for the existence of a faithful representation. We first require a simple lemma on reductive groups.

Lemma 3.25 Let G be an analytic group whose radical is a torus T. Then G is reductive.

Proof. We need to produce a faithful representation of G. Let S be a semisimple subgroup of G with $G = TS$. Then S normalizes, hence centralizes, T, so $T \cap S$ is central in S and therefore, finite. By (3.5), S is an algebraic group and we have an analytic surjection $f: T \times S \to G$ given by $f(t, s) = ts$. Now $T \times S$ is algebraic and $\mathrm{Ker}(f)$ is isomorphic to $T \cap S$ and hence finite, so $(T \times S)/\mathrm{Ker}(f)$ is an algebraic group isomorphic to G, so G has a faithful representation.

Theorem 3.26 Let G be an analytic group. Assume that for every $x \neq e$ in G there is a representation p of G with $p(x) \neq e$. Then G has a faithful representation.

Proof. We are going to show that G has a nucleus by the devices of (3.23). Instead of representations, we will look at homomorphisms of G to algebraic groups with Zariski-dense images.

We first consider the subgroup (G, R) of G. As in (2.3), if $p: G \to H$ is any analytic homomorphism to an algebraic group, we have $(p(G), p(R)) = p(G, R)$ a unipotent subgroup of H. If $K = \mathrm{Ker}(p|(G, R))$ and K_1 is the connected component of e in K, then the discrete group K/K_1 is the kernel of the induced homomorphism from $(G, R)/K_1$ onto $p(G, R)$. The latter being simply-connected implies that $K/K_1 = \{e\}$ so K is connected. Choose any element $x_1 \neq e$ in (G, R) and let $p_1: G \to H_1$ be a homomorphism as above with $p_1(x_1) \neq e$. Then $x_1 \notin K_1 = \mathrm{Ker}(p_1|(G, R))$ and K_1 is a connected analytic subgroup of (G, R). If $K_1 \neq e$, choose $x_2 \neq e$ in K_1. There is a homomorphism $q: G \to H'$, H' algebraic, with $q(x_2) \neq e$. Let $H_2 = H_1 \times H'$ and let $p_2: G \to H_2$ be $p_2(g) = (p_1(g), q(g))$. Then $K_2 = \mathrm{Ker}(p_2|(G, R))$ is contained in

K_1, is connected, and $x_2 \notin K_2$. We can continue in this fashion to produce a descending sequence of connected subgroups K_i of (G, R), with $K_{i+1} \subsetneq K_i$, and homomorphsims $p_i: G \to H_i$ to algebraic groups with $\mathrm{Ker}(p_i|(G, R)) = K_i$. Then $K_n = \{e\}$ for some n, and $p_n: G \to H_n$ is faithful on (G, R), and $\bar{p}_n(G, R)$ is unipotent, so (G, R) is simply-connected.

We next show that $G/(G, R)$ also has the property hypothesized for G. Let $x \in G$, $x \notin (G, R)$. Choose a homomorphism $q: G \to H$, H algebraic, with $q(x) \neq e$. Let $q': G \to H \times H_n$ be given by $q'(g) = (q(g), p_n(g))$, so q' is faithful on (G, R). Assume $q'(x) \in q'(G, R)$. Then there is a unique z in (G, R) with $q'(x) = q'(z)$. Choose $q'': G \to H''$, H'' algebraic, with $q''(xz^{-1}) \neq e$. Let $v: G \to H \times H_n \times H''$ be given by $v(g) = (q'(g), q''(g))$. If $v(x) = v(w)$ with $w \in (G, R)$, we have $q'(x) = q'(w)$ so $w = z$ and $v(xz^{-1}) = e$; but this is impossible since $q''(xz^{-1}) \neq e$. So we have produced a homomorphism v from G to an algebraic group H with $v(x)$ not in $v(G, R)$ and v faithful on (G, R). We can assume that v has Zariski-dense image. Then, as in the proof of (3.23), $v(G, R)$ is Zariski-closed and normal in H, and the induced map $\bar{v}: G/(G, R) \to H/v(G, R)$ is a homomorphism to an algebraic group with $\bar{v}(x(G, R)) \neq e$. We have shown that every nonidentity element of $G/(G, R)$ is nonidentity under some homomorphism from $G/(G, R)$ to an algebraic group.

Now we prove (3.26) under the assumption that $(G, R) = \{e\}$ (as it does in $G/(G, R)$). In this case R is central, hence abelian, so $R = V \times T \times A$, where V is a vector group, T a torus, and A compact analytic. If $p: G \to H$ is a homomorphism to an algebraic group, $p(A) = \{e\}$, so our assumptions on G imply $A = \{e\}$ and $R = V \times T$. Let S be a semisimple subgroup of G with $G = RS$. Then, exactly as in (3.23), TS is a subgroup of G since S centralizes T, and TS is reductive by (3.25), so again following (3.23) we have that V is a nucleus of G.

Finally, we return to our original G. We know that $G/(G, R)$ satisfies the hypothesis of (3.26) by the third paragraph of the proof, hence, has a nucleus by the fourth, and the inverse image of this nucleus, K, is a nucleus of G as in (3.23). (We use the fact that (G, R) is simply-connected.) So G has a faithful representation.

If G is an analytic group with radical R and S is a semisimple subgroup of G with $G = RS$, then S always has a faithful representation, (3.5). This suggests that having a faithful representation for R should be enough to produce one for G. This is indeed the case, as we shall now show.

Theorem 3.27 Let G be an analytic group with radical R. If R has a faithful representation, so does G.

Proof. Again, we want to show that G has a nucleus. Use the faithful representation of R to regard it as a Zariski-dense analytic subgroup of a solvable algebraic group $\bar{R}$. Then $(R, R) = (\bar{R}, \bar{R})$ is simply-connected and R^{ab} is a subgroup of $\bar{R}^{ab}$ so it has a faithful representation. Also (R, R) is closed in $\bar{R}$, hence in R and in G. If $G/(R, R)$ has a nucleus, its inverse image in G will be a nucleus of G, so we may replace G by $G/(R, R)$ and assume R is abelian. Then $R = V \times T$ with V a vector group and T a torus; T is characteristic in R and, hence, normal in G. The adjoint action of G on $L(R)$ factors through G/R, so G acts semisimply on $L(R)$. Since $L(T)$ is G-stable, there is a subspace W of $L(R)$ stable under G with $L(R) = W \oplus L(T)$. Let $K = \exp_R(W)$. Since $\mathrm{Ker}(\exp) \subseteq L(T)$, $R = K \times T$ and K is a vector group. Since K is also normal in G, it follows from (3.25) that G/K is reductive. Thus, K is a nucleus of G and (3.27) is proven.

We can regard (3.26) and (3.27) as applications of (3.7) and (3.16). Of course, (3.16) is really too strong for these applications: if all we wanted to know was that analytic groups with nuclei have faithful representations, we could produce a much simpler argument than (3.16). The full force of (3.16) will become clear in Chapter 4 when we apply the results of this chapter to rings of representative functions.

Summary of results of Chapter 3
An analytic group is *reductive* if it has a faithful representation and if every module is semisimple. A closed normal analytic subgroup K of an analytic group G is a *nucleus* if K is solvable, simply-connected, and G/K is reductive. If K is a nucleus of G,

$G = KH$ with $H \cap K = \{e\}$, where H is a closed reductive subgroup of G. A *split hull* of G is a triple $(\bar{G}, f, T)$ where $\bar{G}$ is an algebraic group, $f: G \to \bar{G}$ is an analytic embedding with Zariski-dense image and T is a torus in $\bar{G}$ with $\bar{G} = f(G)T$ and $T \cap f(G) = \{e\}$. If G has a nucleus then it has a split hull. If $(\bar{G}, f, T)$ is a split hull of G, $h: G' \to \bar{G}$ a homomorphism of algebraic groups and $f': G \to G'$ an analytic homomorphism with Zariski-dense image such that $hf' = f$, then $h^{-1}(T)$ is a torus in G' and $(G', f', f^{-1}(T))$ is a split hull of G. An analytic group with a faithful representation has a nucleus. An analytic group such that the intersection of the kernels of all representations of the group is trivial has a faithful representation.

4

Structure theory of algebras of representative functions

In Chapter 2, we saw that the category Mod(G) of (finite-dimensional, analytic) modules for an analytic group G depends only on the Hopf algebra $R(G)$ of representative functions on G. The finitely-generated Hopf subalgebras of $R(G)$ can be regarded, as we will show below, as coordinate rings of algebraic groups containing G as a Zariski-dense analytic subgroup. This situation was studied in Chapter 3, and we will see in this chapter how the results obtained in Chapter 3 can be used to derive the structure of $R(G)$. In particular, we will show that the group $\mathscr{G}(G)$ of $\mathbb{C}$-algebra homomorphisms from $R(G)$ to $\mathbb{C}$ is a semidirect product of the image of G, which is normal in $\mathscr{G}(G)$, and a group T which is an inverse limit of tori. This result is then used to show that $R(G)$ is, as an algebra, the tensor product over $\mathbb{C}$ of a finitely-generated subalgebra A and a complex group ring $\mathbb{C}[Q]$, where $Q = \exp(\mathrm{Hom}(G, \mathbb{C}))$ and A is such that:

(1) if $f \in A$ and $x \in G$, then $f \cdot x \in A$; and
(2) every maximal ideal of A is the kernel of evaluation at a unique element of G.

The algebra A arises from a split hull of G, and we will show that subalgebras of $R(G)$ like A also give rise to split hulls. As we noted in Chapter 2, knowledge of the algebra structure of $R(G)$ allows, in a sense, a construction of 'generators' of Mod(G), and we will carry this out for the groups of the examples of Chapter 1.

To apply the results of Chapter 3, we need first to reduce to groups with faithful representations.

Theorem 4.1 Let G be an analytic group and let K be the intersection of the kernels of all representations of G. Let $p: G \to G/K$ be the projection. Then G/K has a faithful representation, and $p^*: R(G/K) \to R(G)$ given by $(p^*f)(x) = f(p(x))$ is a (well-defined) Hopf algebra isomorphism.

Proof. To see that G/K has a faithful representation, we use (3.26): let $y \in G/K$, $y \neq e$. Choose $x \in G$ with $p(x) = y$. Then $x \notin K$, so there is a representation $\rho: G \to GL(V)$ with $\rho(x) \neq e$. Let $L = \text{Ker}(\rho)$ and let $\bar{\rho}: G/L \to GL(V)$ be the induced faithful representation, so $\bar{\rho}\rho(xL) = \rho(x) \neq e$. Since $K \subseteq L$, we can consider the composite $\rho': G/K \to G/L \to GL(V)$, and $\rho'(y) = \rho(x) \neq e$. Thus, G/K has a faithful representation. Next, we check that p^* is well-defined. If $f \in R(G/K)$, then $p^*f = fp$ is an analytic function on G. Let $\Delta(f) = \sum h_i \otimes k_i$. Then if $x, y \in G$, $(p^*f)(xy) = f(p(x)p(y)) = \sum h_i(p(x))k_i(p(y)) = \sum (p^*h_i)(x)(p^*k_i)(y)$. By (2.4), then, p^*f is representative. This calculation shows that $\Delta(p^*f) = (p^* \otimes p^*)(\Delta(f))$, and similar elementary calculations show that p^* preserves counit and antipode. Since p^* is obviously linear, p^* is a Hopf algebra homomorphism, and p^* is injective since p is surjective. We still need to see that p^* is surjective. Let f be in $R(G)$. Then there is a G-module V and $v \in V$, $\phi \in V^*$ such that for x in G, $f(x) = \phi(xv)$. Let $\rho: G \to GL(V)$ be the representation associated to V. Then $K \subseteq \text{Ker}(\rho)$, so if $x \in G$ and $y \in K$, $f(xy) = \phi(\rho(x)\rho(y)v) = f(x)$, so f is constant on the cosets of K and, hence, $f = p^*h$ for suitable $h \in R(G/K)$.

Because of (4.1), we will henceforth restrict attention to groups with faithful representations.

Continuing with preliminary considerations, we need to establish some conventional ways of viewing Hopf subalgebras of $R(G)$.

Let G be an analytic group and A a Hopf subalgebra of $R(G)$ finitely-generated over $\mathbb{C}$. Now A is without nilpotents (in fact, A is a domain) since $R(G)$ is a domain, so we can regard A as the coordinate ring of an algebraic group $G(A)$. To keep things definite, the set $G(A)$ is the set of all $\mathbb{C}$-algebra homomorphisms from A to $\mathbb{C}$, with group operations derived from the Hopf algebra operations of A; A is an algebra of functions on $G(A)$, where $f(x) = x(f)$ if $f \in A$ and $x \in G(A)$. We have a canonical homomorphism $e_A: G \to G(A)$ given by $e_A(x)(f) = f(x)$ for $f \in A$ and $x \in G$. Since each element of A is an element of $R(G)$, and hence a function on G, if f in A, regarded as a function on $G(A)$, vanishes on $e_A(G)$, it is zero on $G(A)$. Thus, $e_A(G)$ is Zariski-dense in $G(A)$. Also e_A is analytic: if $G(A) \to \mathrm{GL}_n \mathbb{C}$ is any algebraic embedding, then the matrix coefficient functions on $\mathrm{GL}_n \mathbb{C}$ restrict to polynomial functions on G, i.e., to elements of A. If $f \in A$ is one of these functions, then $(fe_A)(x) = (e_A(x))(f) = f(x)$ for x in G, so fe_A is analytic on G.

In summary: a finitely-generated Hopf subalgebra A of $R(G)$ gives rise to an algebraic group $G(A)$ and an analytic homomorphism $e_A: G \to G(A)$ with Zariski-dense image.

Conversely, suppose H is an algebraic group and $d: G \to H$ is an analytic homomorphism with Zariski-dense image. Choose an algebraic embedding $c: H \to \mathrm{GL}(V)$, for suitable vector space V, and let $\rho: G \to \mathrm{GL}(V)$ be the composite. In (2.29) we saw that ρ gives rise to a Hopf algebra homomorphism $\rho^*: \mathbb{C}[\mathrm{GL}(V)] \to R(G)$ where $\rho^*(h)(x) = h(\rho(x))$ for $h \in \mathbb{C}[\mathrm{GL}(V)]$ and $x \in G$. Now the Zariski-closure of $\rho(G)$ in $\mathrm{GL}(V)$ is $c(H)$, and it follows from (2.30) that ρ^* factors as $\mathbb{C}[\mathrm{GL}(V)] \to \mathbb{C}[c(H)] \to R(G)$ where the second map is an injection (the first is, of course, given by restriction of functions). Thus, the Hopf algebra $A = \mathrm{Image}(\rho^*)$ is isomorphic to $\mathbb{C}[c(H)]$ and, hence, to $\mathbb{C}[H]$. To be explicit, the isomorphism $\mathbb{C}[H] \to A$ sends f in $\mathbb{C}[H]$ to $\rho^*(g)$, where g is any element of $\mathbb{C}[\mathrm{GL}(V)]$ which restricts to fc^{-1} on $c(H)$. Now for x in G, $\rho^*(g)(x) = g(\rho(x)) = g(cd(x)) = f(d(x))$, so the isomorphism is just right composition with d, which we will denote d^*. Now $d^*: \mathbb{C}[H] \to A$ gives an algebraic group isomorphism $\phi: G(A) \to H$. We want to compute the composite ϕe_A. So let $x \in G$. Then $\phi(e_A(x))$ is the unique element of H such that, for all f in $\mathbb{C}[H]$, $f(\phi(e_A(x))) = d^*(f)(e_A(x))$. Since $d^*(f)(e_A(x)) = d^*(f)(x) = f(d(x))$, $\phi e_A(x) = d(x)$ and, hence, $\phi e_A = d$.

In summary: if H is an algebraic group and $d: G \to H$ an analytic homomorphism with Zariski-dense image, then there is a Hopf algebra injection $d^*: \mathbb{C}[H] \to R(G)$ given by $d^*(f) = fd$. If A is the image of d^* there is an algebraic group isomorphism $\phi: G(A) \to H$ such that $\phi e_A = d$.

The above discussion shows that the finitely-generated Hopf subalgebras of $R(G)$ lead to algebraic groups to which G maps analytically with Zariski-dense image, and moreover, that every such homomorphism to an algebraic group can be realized inside $R(G)$. (This observation should be compared to (2.14), which shows how G-modules can be realized as modules of representative functions.) In particular, this means that if G has a faithful representation, we can realize a split hull of G in $R(G)$. We need to analyze some of the consequences of the presence of these split hulls, and especially of their consequences in terms of characters of G.

Definition 4.2   Let G be an analytic group. A character $f: G \to \mathbb{C}^*$ is *exponential additive* if there is an additive character $\phi: G \to \mathbb{C}$ such that $f(x) = e^{\phi(x)}$ for all $x \in G$.

Proposition 4.3   Let G be an analytic group, let B be a finitely-generated Hopf subalgebra of $R(G)$ and suppose there is a torus T in B such that $(G(B), e_B, T)$ is a split hull of G. Then there is a unique algebraic group homomorphism $d: G(B) \to T$ such that
  (i) $d(x) = x$ for $x \in T$;
  (ii) if $x \in G$ and $e_B(x)(G(B), G(B))$ is in the maximal torus of $G(B)^{ab}$, $d(e_B(x)) = e$.
Moreover, $X(T)d$ is the set of exponential additive characters of G in B.

Proof.   Let $\bar{H} = G(B)^{ab}$ and let $p: G(B) \to \bar{H}$ be the projection. Let $\bar{G} = p(e_B(G))$ and let $\bar{T} = p(T)$. By (3.1), $\bar{G}$ is isomorphic to G^{ab} and $\bar{H} = \bar{G} \cdot \bar{T}$ with $\bar{T} \cap \bar{G} = \{e\}$. Now $\bar{G}$, as an analytic subgroup of the algebraic group $\bar{H}$, has a faithful representation and, hence, a nucleus, so we can write $\bar{G} = K \cdot P$ where K is a nucleus of $\bar{G}$, P is reductive and $P \cap K = \{e\}$ by (3.7). Since $\bar{G}$ is abelian, K is a vector group and P is a torus. Let U be the unipotent radical of $\bar{H}$ and let S be the (unique) maximal torus. Then $S = \operatorname{Ker}(X^+(H))$. But also $\bar{H} = \bar{G} \cdot \bar{T} = K \times P \times \bar{T}$, so $P \cdot \bar{T} = \operatorname{Ker}(X^+(H))$ also. Thus, $S = P \cdot \bar{T}$ with $P \cap \bar{T} = \{e\}$. Under the algebraic group homomorphism $H \to \bar{H} \to \bar{H}/U \cdot P$, T maps isomorphically to $\bar{T}$ which maps isomorphically to the quotient. We define $d: H \to T$ to be $H \to \bar{H}/U \cdot P$ followed by the inverse of the isomorphism $T \to \bar{H}/U \cdot P$. We clearly have that d satisfies condition (i).

To see that d satisfies condition (ii), we will need that $\bar{G} \cap S = P$. By construction, $P \subseteq \bar{G} \cap S$. To see that the converse inclusion holds, we consider the algebraic group homomorphism $q: \bar{H} \to U$ which is the identity on U (the projection). Since q is algebraic and $\bar{G}$ is Zariski-dense in $\bar{H}$, $q(\bar{G})$ is Zariski-dense in U and since U is unipotent $q(\bar{G}) = U$ by $(3.1)(c)$. Since $\mathrm{Ker}(q) = S$ and $P \subseteq S$, $q(P) = \{e\}$, so $U = q(\bar{G}) = q(KP) = q(K)$. Now $\bar{H} = K \cdot P \cdot T = K \times S = U \times S$ so $\dim(K) = \dim(U)$ and, hence, $(q\,|K): K \to U$ is an isomorphism (remember, K and U are vector groups). Thus, $K \cap S = K \cap \mathrm{Ker}(q) = \{e\}$. If $x \in \bar{G} \cap S$ and $x = kp$ with $k \in K$ and $p \in P$, then $p \in G \cap S = \{e\}$ so $k \in \bar{G} \cap S$ and, hence, $k \in S \cap K = \{e\}$, so $x = p \in P$. Thus, $\bar{G} \cap S \subseteq P$, so $\bar{G} \cap S = P$.

Now we verify condition (ii): d is constructed to vanish on $p^{-1}(\bar{G} \cap S) = e_B(G) \cap p^{-1}(S)$, and if $x \in G$ is such that $e_B(x) \times (G(B), G(B)) \in S$, then $p(e_B(x)) \in S$ so $e_B(x)$ is in $e_B(G) \cap p^{-1}(S)$ and, hence, $d(e_B(x)) = e$.

Next we check that (i) and (ii) characterize d. Suppose $d': G(B) \to T$ is an algebraic group homomorphism satisfying (i) and (ii). Then d' induces an algebraic group homomorphism $\bar{d}': \bar{H} \to T$ vanishing on $\bar{G} \cap S = P$. Of course, $\bar{d}'$ must also vanish on U. If $x \in \bar{H}$, write $x = uyp(t)$ where $u \in U$, $y \in P$ and $t \in T$ (this can be done in a unique fashion). Then $\bar{d}'(x) = \bar{d}'(p(t)) = d'(t) = t$. Thus, $\bar{d}'$, and hence d, is uniquely determined by the conditions (i) and (ii), so $d' = d$.

Finally, we need to verify that if $f \in X(T)$, fd is an exponential additive character of G in B, and conversely. Of course, $fd \in B = \mathbb{C}[G(B)]$. The algebraic group morphism d factors through $\bar{H}$: if $\bar{d}: \bar{H} \to T$ is given by $\bar{d}(uyp(t)) = t$ for $u \in U, y \in P, t \in T$, then $d = \bar{d}p$. Since $\bar{d}(P) = \{e\}$, $\bar{d}|\bar{G}$ factors through the projection of $\bar{G} = K \cdot P$ onto K. This means that, as a function on G, $fd = f\bar{d}p$ can be written as $G \to \bar{G} \to K \to \mathbb{C}^*$ and, since K is a vector group, the final map $K \to \mathbb{C}^*$ factors as $K \to \mathbb{C} \to \mathbb{C}^*$ where the first map is linear and the second is the exponential. Thus, fd factors as $G \to \mathbb{C} \to \mathbb{C}^*$ which shows that fd is exponential additive.

Conversely, suppose $g \in B$ is an exponential additive character of G. Then g is actually an algebraic character of $G(B)$: for if $x, y \in G$, $g(xy) = g(x)g(y)$, so $\Delta(g) = g \otimes g$ and $I(g) = g^{-1}$, so

$\mathbb{C}[GL_1(\mathbb{C})] = \mathbb{C}[t, t^{-1}] \to B$ by $t \mapsto g$ is a Hopf algebra homomorphism which leads to an algebraic character $h: G(B) \to \mathbb{C}^*$ with $he_B = g$. Then h induces an algebraic character $\bar{h}: \bar{H} \to \mathbb{C}^*$ such that $g = \bar{h}pe_B$. Now $P' = e_B^{-1}(e_B(G) \cap p^{-1}(S)) = e_B^{-1}(p^{-1}(P))$ contains (G, G) and $P'/(G, G)$ is isomorphic to the torus P. Since g factors as $G \to G^{ab} \to \mathbb{C} \to \mathbb{C}^*$, $g(P') = \{e\}$, so $\bar{h}(P) = \{e\}$. Since $\bar{h}$ is algebraic, $\bar{h}(U) = \{e\}$ also. This means that $\bar{h}$ factors as $\bar{H} \to T \to \mathbb{C}^*$ where the first map is $\bar{d}$ and the second is some character f of T. Thus, $g = f\bar{d}pe_B = fde_B$ is in $X(T)d$.

We also need a relative version of (4.3) where we allow B to be enlarged.

Lemma 4.4 Let G be an analytic group, let B be a finitely-generated Hopf subalgebra of $R(G)$ and suppose T is a torus in $G(B)$ such that $(G(B), e_B, T)$ is a split hull of G. Let B' be a finitely-generated Hopf subalgebra of $R(G)$ containing B and let $f: G(B') \to G(B)$ be the algebraic group homomorphism induced from the inclusion $B \subseteq B'$. Let $T' = f^{-1}(T)$. Then $(G(B'), e_{B'}, T')$ is a split hull of G and
  (i) $\{f \in B \,|\, t \cdot f = f$ for all $t \in T\} = \{f \in B' \,|\, t \cdot f = f$ for all $t \in T'\}$;
  (ii) if $d: G(B) \to T$ and $d': G(B) \to T'$ are the homomorphisms given by (4.3), then $df = (f|T')d'$.

Proof. $(G(B'), e_{B'}, T')$ is a split hull by (3.17). Let $\bar{e}_B: G \to G(B)/T$ be $\bar{e}_B(x) = e_B(x)T$ and $\bar{e}_{B'}: G \to G(B')/T'$ be $\bar{e}_{B'}(x) = e_{B'}(x)T'$. Since $G(B)$ and $G(B')$ are split hulls, $\bar{e}_B$ and $\bar{e}_{B'}$ are bijections. Now f induces a surjection $\bar{f}: G(B')/T' \to G(B)/T$, and $\bar{f}\bar{e}_{B'} = \bar{e}_B$, so $\bar{f}$ is an isomorphism of algebraic varieties. Thus, $\bar{f}$ determines an isomorphism of coordinate rings $\mathbb{C}[G(B)/T] \to \mathbb{C}[G(B')/T']$. Now $\mathbb{C}[G(B)/T]$ is just the left-hand side of (i) while $\mathbb{C}[G(B')/T']$ is the right-hand side and the isomorphism induced from $\bar{f}$ is the inclusion so (i) holds.

To establish (ii), let $L = \mathrm{Ker}(f)$. Then $L \subseteq T'$, so $(f|T')d': G(B') \to T$ induces an algebraic group homomorphism $g: G(B) \to T$, such that $gf = (f|T')d'$. If $x \in T$ and $y \in T'$ with $f(y) = x$, then $g(x) = gf(y) = fd'(y) = f(y) = x$, so g satisfies (i) of (4.3). Suppose x in G is such that $e_B(x)(G(B), G(B))$ is in the maximal torus of S of

$G(B)^{ab}$. Now f induces a surjection $\bar{f}: G(B')^{ab} \to G(B)^{ab}$ which carries the maximal torus S' of $G(B')^{ab}$ onto S and carries the image $\bar{G}'$ of $e_{B'}(G)$ bijectively to the image $\bar{G}$ of $e_B(G)$ (since $fe_{B'} = e_B$). It follows that $\bar{f}(\bar{G}' \cap S') = \bar{G} \cap S$ and, hence, that $e_{B'}(x)(G(B'), G(B'))$ is in S'. Thus, $d'(e_{B'}(x)) = e$, so $g(e_B(x)) = gfe_{B'}(x) = (f|T')d'(e_B(x)) = e$ and g satisfies (ii) of (4.3). By (4.3), $g = d$ so $df = (f|T')d'$.

We also want to interpret (4.3) in terms of the ring B.

Proposition 4.5 Let G be an analytic group, let B be a finitely-generated Hopf subalgebra of $R(G)$ and let T be a torus in $G(B)$ such that $(G(B), e_B, T)$ is a split hull of G. Let $D = \{f \in B \mid t \cdot f = f$ for all $t \in T\}$ and let $d: G(B) \to T$ be the morphism of (4.3). Then $D \otimes_{\mathbb{C}} \mathbb{C}[T] \to B$ by $a \otimes f \mapsto a(fd)$ is a $\mathbb{C}$-algebra isomorphism, and B is the group ring $D[Q_B]$ where Q_B is the group of exponential additive characters of G in B.

Proof. Let $p: G(B) \to G(B)/T$ be the canonical projection. Then $G(B) \to (G(B)/T) \times T$ by $x \mapsto (xT, d(x))$ is an isomorphism of varieties. (The inverse sends (xT, t) to $xd(x)^{-1}t$.) Thus, $G(B)/T$ is affine and p induces an injection $\mathbb{C}[G(B)/T] \to B$ whose image is D. Since $\mathbb{C}[(G(B)/T) \times T] = \mathbb{C}[G(B)/T] \otimes \mathbb{C}[T]$, the first assertion follows. Since $\mathbb{C}[T]$ is the group ring $\mathbb{C}[X(T)]$ and the map $\mathbb{C}[T] \to B$ induced from d sends $X(T)$ to Q_B by (4.3), the second is also established.

We can extend the description of B in (4.5) to a similar description of $R(G)$, for $R(G)$ is the direct limit of its finitely-generated Hopf subalgebras; if we begin the direct limit with B then (4.4) and (4.5) will allow us to show that $R(G)$ is the group ring $A[Q]$ where Q is the group of exponential additive characters of G. This same type of direct limit argument will be used to describe the group $\mathscr{G}(G)$ defined in (2.31), so we will introduce the appropriate notations for the direct limit first.

Notation 4.6 Let G be an analytic group and let $\{A_i \mid i \in I\}$ be the set of finitely-generated Hopf subalgebras of $R(G)$. Order I so that $i \le j$ if $A_i \subseteq A_j$. For $i \in I$, let $G_i = G(A_i)$ and let $e_i = e_{A_i}: G \to G_i$.

If $i, j \in I$ with $i \le j$, let $f_{ij}: G_j \to G_i$ be the algebraic group homomorphism induced by the inclusion $A_i \subseteq A_j$ (so if $x: A_j \to \mathbb{C}$ is in G_j, then $f_{ij}(x): A_i \to \mathbb{C}$ is the restriction of x to A_i); I is directed, and $R(G) = \bigcup \{A_i \,|\, i \in I\} = \mathrm{dir} \lim \{A_i \,|\, i \in I\}$.

Theorem 4.7 Let G be an analytic group with a faithful representation and let Q be the group of exponential additive characters of G. Let B be a finitely-generated Hopf subalgebra of $R(G)$ containing a torus T such that $(G(B), e_B, T)$ is a split hull of G. Let $D = \{f \in B \,|\, t \cdot f = f$ for all $t \in T\}$. Then (as algebra) $R(G)$ is the group ring $D[Q]$.

Proof. We follow Notation 4.6. There is $k \in I$ such that $B = A_k$. Let $I' = \{i \in I \,|\, k \le i\}$. Then, $R(G) = \bigcup \{A_i \,|\, i \in I'\}$, since I is directed. For $i \in I'$, let $T_i = f_{ki}^{-1}(T)$. Then (G_i, e_i, T_i) is a split hull of G (3.17). Let $D_i = \{f \in A_i \,|\, t \cdot f = f$ for all $t \in T_i\}$ and let $Q_i = Q \cap A_i$. By (4.5), A_i is the group ring $D_i[Q_i]$, so Q_i freely generates A_i as a D_i-module. By (4.4)(i), $D_i = D$, so $A_i = D[Q_i]$. But $R(G) = \bigcup \{A_i \,|\, i \in I'\}$ so $R(G) = \bigcup D[Q_i] = D[Q]$.

Theorem 4.8 Let G be an analytic group with a faithful representation. Then the canonical homomorphism (2.31) $\tau: G \to \mathscr{G}(G)$ is injective, $\tau(G)$ is normal in $\mathscr{G}(G)$ and there is a subgroup T of $\mathscr{G}(G)$ such that $\mathscr{G}(G) = \tau(G) \cdot T$ with $\tau(G) \cap T = \{e\}$; T is the inverse limit of algebraic tori, and as an abstract group T is isomorphic to $\mathrm{Hom}(X^+(G), \mathbb{C}^*)$.

Proof. We follow notation in (4.6). There is a $k \in I$ and a torus T_k in G_k such that (G_k, e_k, T_k) is a split hull of G. Let $I' = \{i \in I \,|\, k \le i\}$. We first note that the directed inverse system $\{G_i, f_{ij}: G_j \to G_i \,|\, i \in I'\}$ and the analytic maps $e_i: G \to G_i$ satisfy the hypotheses of (3.21), with $i_0 = k$. Now $f_{ij}e_j = e_i$ for $i, j \in I'$ with $i \le j$, so (3.21) does apply. Let H be the inverse limit of the system $\{G_i, f_{ij} \,|\, i \in I'\}$; $\mathscr{G}(G)$ is the set of $\mathbb{C}$-algebra homomorphisms from $R(G)$ to $\mathbb{C}$, G_i is the set of $\mathbb{C}$-algebra homomorphisms from A_i to $\mathbb{C}$ and $R(G) = \mathrm{dir} \lim \{A_i \,|\, i \in I'\}$. Thus, $\mathscr{G}(G)$ can be identified with the inverse limit H. Under this identification, the map $G \to \mathscr{G}(G)$ induced from the maps $e_i: G \to G_i$ is just τ, for the projection $f_i: \mathscr{G}(G) \to G_i$ is just restriction of algebra homomorphisms from $R(G)$ to A_i and, hence, $\tau f_i = e_i$ for all i. From (3.21) we conclude all the statements of the theorem except for the description of T as an abstract group.

For this description, we must look at the proof of (3.21). For $i \in I'$, let $T_i = f_{ki}^{-1}(T_k)$. Then (G_i, e_i, T_i) is a split hull of G; let $d_i : G_i \to T_i$ be the unique algebraic group homomorphism of (4.3). The proof of (3.21) (which includes (3.22) and the subsequent remark) shows that T is the group of all $\mathbb{C}$-algebra homomorphisms from the complex group ring $\mathbb{C}[\text{dir} \lim X(T_i)]$ to $\mathbb{C}$. This is just $\text{Hom}(\text{dir} \lim(X(T_i)), \mathbb{C}^*)$, so we must determine $\text{dir} \lim(X(T_i))$. In the direct system $\{X(T_i) \mid i \in I'\}$ the transition map $X(T_i) \to X(T_j)$ is induced from $f_{ij} \mid T_j : T_j \to T_i$ and sends g to $g(f_{ij} \mid T_j)$. Let Q be the group of exponential additive characters of G and let $Q_i = Q \cap A_i$. Then $Q = \text{dir} \lim\{Q_i \mid i \in I'\}$, where the transition map $Q_i \to Q_j$ is inclusion. By (4.3), d_i induces a group isomorphism $d_i^* : X(T_i) \to Q_i$ by $d_i^*(g) = gd_i$. We claim that $\{d_i^* \mid i \in I'\} : \{X(T_i) \mid i \in I'\} \to \{Q_i \mid i \in I'\}$ is an isomorphism of inverse systems: this means that if $g \in X(T_i)$, then $d_j^*(g(f_{ij} \mid T_j)) = d_i^*(g)$, or that $g(f_{ij} \mid T_j)d_j = gd_i$. But by (4.4)(ii), $(f_{ij} \mid T_j)d_j = d_i f_{ij}$, so $g(f_{ij} \mid T_j)d_j = gd_i f_{ij}$. Since $gd_i \in A_i$ and $A_i \to A_j$ by $h \mapsto hf_{ij}$ is just inclusion, this means that $gd_i f_{ij} = gd_i$ so we do indeed have an isomorphism of inverse systems. Thus, $\text{dir} \lim\{X(T_i) \mid i \in I'\} = Q$ and $T = \text{Hom}(Q, \mathbb{C}^*)$. Since $X^+(G)$ maps isomorphically to Q by $\phi \mapsto e^\phi$, the desired description of T results.

Theorems 4.7 and 4.8 yield much information about the nature of the category of G-modules. We are going to examine their consequences in detail in the rest of this chapter.

It is important to notice that (4.7) gives the structure of $R(G)$ only as an algebra, not as a Hopf algebra. However, as we saw in (2.15), knowledge of a set of algebra generators $\{f_i \mid i \in I\}$ for $R(G)$ gives a uniform method to generate G-modules starting with the set of G-modules $\{[f_i]_r \mid i \in I\}$. (The reason that the modules $[f_i]_r$ rather than $_l[f_i]$ were chosen will become clear shortly.) We can use (4.7) to produce a set of algebra generators for $R(G)$: let $S = \{f_i \mid i \in I\}$ be a set of algebra generators for D and then $S \cup Q$ is a set of algebra generators for $R(G)$. If $a \in Q$ and $x \in G$, then $a \cdot x = a(x)a$ so $[a]_r = \langle a \rangle$, while if $f \in S$, $[f]_r = \langle f \cdot x \mid x \in G \rangle$. Now if $t \in T$ and $x \in G$, $t \cdot (f \cdot x) = (t \cdot f) \cdot x = f \cdot x$, so $[f]_r \subseteq D$. (This is the reason we work on the right.) We also need to observe, for example, using (4.5), that D is a finitely-generated algebra, so we can take S to be finite. These considerations can be summed up by saying that all G-modules can be generated from the one-dimensional modules coming from exponential additive characters, along with a finite set of G-submodules of D. To undertake this construction for a particular group G, we only need produce D, which means to construct a split hull of G. We show how the examples of Chapter 1 can be interpreted in this light.

In dealing with the modules $[f]_r$ for f in $R(G)$ we must remember how G acts: if $x \in G$ and $v \in [f]_r$, then $x(v) = v \cdot x^{-1}$. Note the inverse; this means, for example, if f is a character, that although $[f]_r = \langle f \rangle$, we have $x(f) = f(x^{-1})f = f^{-1}(x)f$, so $[f]_r$ is a one-dimensional module corresponding to the character f^{-1}.

Example A″ Let $G = \mathrm{GL}_1 \mathbb{C}$. Then G is an algebraic group, so $(G, \mathrm{id}, \{e\})$ is a split hull of G. The Hopf algebra injection $(\mathrm{id})^* : \mathbb{C}[G] \to R(G)$ sends the coordinate function t on G (given by $t(x) = x$) to t regarded as a representative function on G. Since G has no additive characters, it has no exponential additive characters, so $(\mathrm{id})^*$ is an isomorphism. In the present circumstances $D = R(G)$ and t, t^{-1} are the algebra generators of D. Thus, the one-dimensional G-modules $[t]_r = \langle t \rangle$ and $[t^{-1}]_r = \langle t^{-1} \rangle$ generate the category $\mathrm{Mod}(G)$. (We note that these are, of course, the same answers as obtained in Examples A and A′.)

Example B″ $G = (\mathbb{C}, +)$. Once again, G is algebraic and $(G, \mathrm{id}, \{e\})$ is a split hull of G. Let t be the coordinate function on G given by $t(x) = x$. Then $\mathbb{C}[G] = \mathbb{C}[t]$ and the image of the Hopf algebra injection $(\mathrm{id})^* : \mathbb{C}[G] \to R(G)$ is $\mathbb{C}[t]$ (here t is regarded as a representative function on G). In the notation of (4.7), $D = \mathbb{C}[t]$ (since $T = \{e\}$), and t is an algebra generator of D. The corresponding G-module $[t]_r = \langle 1, t \rangle$ will be denoted V_2. Now G has additive characters, in fact, $X^+(G) = \mathbb{C}$ and, for each $a \in \mathbb{C}$, let $V(a)$ be the G-module corresponding to the exponential additive character e^{-at}, so $V(a) = [e^{-at}]_r = \langle e^{-at} \rangle$. Then V_2 and $\{V(a) \mid a \in \mathbb{C}\}$ generate

Mod(G), and $R(G) = D[Q] = \mathbb{C}[t][\{e^{at} \mid a \in \mathbb{C}\}]$. (Again, this agrees with the previous calculations in Examples B and B'.)

Example C″ G is the semidirect product of $\mathbb{C}$ and $\mathbb{C}^*$, with $\mathbb{C}^*$ acting on $\mathbb{C}$ as $GL_1\mathbb{C}$. As in the first two examples, G is an algebraic group. Moreover, $X^+(G) = \{e\}$. As in Example A″, this means $R(G)$ is isomorphic to $\mathbb{C}[G]$. Now let $x: G \to \mathbb{C}$, $t: G \to \mathbb{C}$ be the first and second coordinate projections. Then $R(G) = \mathbb{C}[x, t, t^{-1}]$. For $(a, s) \in G$, we have $x \cdot (a, s) = a + sx$ and $t \cdot (a, s) = st$. So $[x]_r = \langle 1, x \rangle$ and $[t]_r = \langle t \rangle$. These two modules generate Mod(G), and it is elementary to check that $\langle 1, x \rangle$ is isomorphic to W_2 and $\langle t \rangle$ is isomorphic to W_1 in the notations of Example C, where we first saw this generation explicitly.

Example D″ G is the semidirect product of $\mathbb{C}$ with $\mathbb{C}$, where the latter acts on the former via the exponential. This group is definitely not algebraic, so we must construct a split hull to apply (4.7). Lemma 3.13 gives a procedure for this construction (in the notation of (3.13) take $N = \mathbb{C} \times \{0\}$ and $C = \{0\} \times \mathbb{C}$), but we will skip to the result, leaving the task of showing that this is where the procedure leads to the interested reader. Thus, we let $\bar{G}$ be the algebraic group of all triples (x, y, t) with $x, y \in \mathbb{C}$ and $t \in \mathbb{C}^*$ and multiplication $(x, y, t)(x', y', t') = (x + tx', y + y', tt')$. Then $f: G \to \bar{G}$ by $f((a, b)) = (a, b, e^b)$ is an analytic injection with Zariski-dense image, and if $T = \{(a, b, t) \in \bar{G} \mid a = b = 0\}$, then $(\bar{G}, f, T)$ is a split hull for G. We write $\mathbb{C}[\bar{G}] = \mathbb{C}[x, y, t, t^{-1}]$ where x, y, t are the first, second and third coordinate functions on $\bar{G}$. The image of the Hopf algebra injection $\mathbb{C}[\bar{G}] \to R(G)$ is the algebra generated by functions xf, yf and tf, whose values on the element (a, b) of G are a, b and e^b, respectively. Then D will be the image of $D' = \{h \in \mathbb{C}[\bar{G}] \mid s \cdot h = h$ for all $s \in T\}$ (this is a slight departure from the notation of (4.7) to avoid needless notational complexity). For $s = (0, 0, r) \in T$, $s \cdot x = x$, $s \cdot y = y$ and $s \cdot t = rt$. If we regard $\mathbb{C}[\bar{G}]$ as $(\mathbb{C}[x, y])[t, t^{-1}]$ and note that $\mathbb{C}[x, y] \subseteq D'$, it is clear that $\mathbb{C}[x, y] = D'$. Thus, $D = \mathbb{C}[xf, yf]$. Let $a = xf$ and $b = yf$. Then a and b generate D as an algebra and $[a]_r = \langle 1, a \rangle$ and $[b]_r = \langle 1, b \rangle$. Now $X^+(G) = \mathbb{C}$, so for each $d \in \mathbb{C}$, we have the G-module $V(d)$ corresponding to the exponential additive character e^{-db}, so $V(d) = [e^{-db}]_r = \langle e^{-db} \rangle$. We want to check that this set of generators is the same as that produced in Example D. The module $\langle 1, b \rangle = [b]_r$ has basis $y_0 = b$ and $y_1 = 1$, where $(c, d)(y_0) = b \cdot (c, d)^{-1} = b \cdot (-e^{-d}c, -d) = -d + b = y_0 + dy_1$ while $(c, d)(y_1) = y_1$, so $[b]_r$ is isomorphic to the module W_2 of Example D. The module $\langle 1, a \rangle = [a]_r$ has basis $x_0 = a$ and $x_1 = -1$ where $(c, d)(x_0) = a \cdot (-e^{-d}c, -d) = -e^{-d}c + e^{-d}a = e^{-d}(x_0 + cx_1)$ and $(c, d)(x_1) = x_1$. The module $V(1)$ has basis $z = e^{-b}$ where $(c, d)z = e^{-b}(c, d)^{-1} = e^d z$, so $[a]_r \otimes V(1)$ has basis $u_0 = x_0 \otimes z$ and $u_1 = x_1 \otimes z$ where $(c, d)(u_0) = e^{-d}(x_0 + cx_1) \otimes e^d z = u_0 + cu_1$ and $(c, d)u_1 = x_1 \otimes e^d z = e^d u_1$. Thus, $[a]_r \otimes V(1)$ is isomorphic to the module V_2 of Example D. Thus, our new generators $[a]_r$, $[b]_r$ and $\{V(d) | d \in \mathbb{C}\}$ produce the old ones V_2, W_2 and $\{V(d) | d \in \mathbb{C}\}$ and vice versa.

Example E″   Let $\alpha, \beta \in \mathbb{C}$ and let $G_{\alpha,\beta}$ be the semidirect product of $\mathbb{C}^{(2)}$ with $\mathbb{C}$, where $\mathbb{C}$ acts on $\mathbb{C}^{(2)}$ via $s : \mathbb{C} \to \mathrm{GL}_2\mathbb{C}$ where $s(x) = \mathrm{diag}(e^{\alpha x}, e^{\beta x})$. For convenience, we assume α, β are nonzero. Let $\bar{G}$ be the algebraic group of quintuples (a, b, x, s, t) where $a, b, x \in \mathbb{C}$ and $s, t \in \mathbb{C}^*$, with group operation $(a, b, x, s, t)(a', b', x', s', t') = (a + sa', b + tb', x + x', ss', tt')$. Let $T = (0, 0, 0) \times (\mathbb{C}^*)^{(2)}$, which is a torus in $\bar{G}$, and define $f_{\alpha,\beta} : G_{\alpha,\beta} \to \bar{G}$ by $f_{\alpha,\beta}(a, b, x) = (a, b, x, e^{\alpha x}, e^{\beta x})$. Then $(\bar{G}, f_{\alpha,\beta}, T)$ is a split hull of $G_{\alpha,\beta}$. It follows that the subalgebra D of $R(G_{\alpha,\beta})$ of (4.7) is the polynomial algebra in the three coordinate functions of $G_{\alpha,\beta}$ (which we denote a, b, x), and since $X^+(G) = \mathbb{C}$ (every additive character of $G_{\alpha,\beta}$ is of the form $(a, b, x) \mapsto \lambda x$ for some $\lambda \in \mathbb{C}$), we have $R(G_{\alpha,\beta}) = \mathbb{C}[a, b, x] \times [Q]$ where $Q = \{e^{\lambda x} | \lambda \in \mathbb{C}\}$.

This is the algebra description of $R(G_{\alpha,\beta})$. We also need to know the Hopf algebra structure. If g, g' are in $G_{\alpha,\beta}$, then $a(gg') = a(g) + e^{\alpha x(g)}a(g')$, so $\Delta(a) = a \otimes 1 + e^{\alpha x} \otimes a$. Similarly, we find $\Delta(b) = b \otimes 1 + e^{\beta x} \otimes b$, and $\Delta(e^{\lambda x}) = e^{\lambda x} \otimes e^{\lambda x}$. Now let $h : \mathbb{C} \to \mathbb{C}$ be an additive group isomorphism and let $\alpha' = h(\alpha)$, $\beta' = h(\beta)$. Then $R(G_{\alpha',\beta'}) = \mathbb{C}[a', b', x'][Q']$ where $Q' = \{e^{\lambda x'} | \lambda \in \mathbb{C}\}$ and a', b', x' are the coordinate functions on $G_{\alpha',\beta'}$. We have an algebra isomorphism $h_* : R(G_{\alpha,\beta}) \to R(G_{\alpha',\beta'})$ defined by $h_*(a) = a', h_*(b) = b', h_*(x) = x'$ and $h_*(e^{\lambda x}) = e^{\lambda x'}$, which is also a Hopf algebra isomorphism.

Since the Hopf algebras $R(G_{\alpha,\beta})$ and $R(G_{\alpha',\beta'})$ are isomorphic, so are the categories $\text{Mod}(G_{\alpha,\beta})$ and $\text{Mod}(G_{\alpha',\beta'})$, as we saw in Example E.

In general, (4.7) only gives the structure of $R(G)$ as an algebra, as we have already noted. If G is an algebraic group, however, (4.7) is actually a Hopf algebra description.

Proposition 4.9 Let G be an algebraic group regarded as an analytic group. Then $R(G)$ is isomorphic as a Hopf algebra to the group algebra $\mathbb{C}[G][X^+(G)]$.

Proof. $(G, \text{id}, \{e\})$ is a split hull of G. Let B be the image of $\mathbb{C}[G]$ in $R(G)$ under the Hopf algebra injection $(\text{id})^*$. Then, $(G(B), e_B, \{e\})$ is a split hull of G, so by (4.7) $R(G) = B[Q]$ as algebras. If $b \in B$, $\Delta(b)$ is in $B \otimes B$, since B is a Hopf subalgebra of $R(G)$, and if $q \in Q$, $\Delta(q) = q \otimes q$. It follows that $\mathbb{C}[G][X^+(G)] \to R(G)$ by $\sum a_i \phi_i \mapsto \sum (\text{id})^* a_i \, e^{\phi_i}$ is a Hopf algebra isomorphism.

Corollary 4.10 Let G be an algebraic group regarded as an analytic group. Then $\mathscr{G}(G)$ is isomorphic to $G \times \text{Hom}(X^+(G), \mathbb{C}^*)$.

Proof. By (4.9), $R(G)$ is isomorphic to the tensor product $\mathbb{C}[G] \otimes \mathbb{C}[X^+(G)]$ as Hopf algebras. Now $\mathscr{G}(G)$ is the set of $\mathbb{C}$-algebra homomorphisms from $R(G)$ to $\mathbb{C}$, while the $\mathbb{C}$-algebra homomorphisms from $\mathbb{C}[G]$ to $\mathbb{C}$ are identified with G and those from $\mathbb{C}[X^+(G)]$ to $\mathbb{C}$ are identified with $\text{Hom}(X^+(G), \mathbb{C}^*)$. Thus, as a set, $\mathscr{G}(G)$ equals $G \times \text{Hom}(X^+(G), \mathbb{C}^*)$ and the fact that we have a Hopf algebra isomorphism makes this description of $\mathscr{G}(G)$ a group isomorphism.

Quite generally, (4.8) shows that the departure of $\mathscr{G}(G)$ from G depends on the nontriviality of $X^+(G)$. We record what happens when $X^+(G)$ is trivial.

Proposition 4.11 Let G be an analytic group with a faithful representation. Then $\tau: G \to \mathscr{G}(G)$ is an isomorphism if and only if $X^+(G) = \{e\}$. If τ is an isomorphism, $\mathscr{G}(G)$ is an algebraic group with coordinate ring $\mathbb{C}[\mathscr{G}(G)] = R(G)$, and the category $\text{Mod}(G)$ is isomorphic to the category of algebraic $\mathscr{G}(G)$-modules.

Proof. The first assertion follows directly from (4.8): in the notation of that theorem, $T = \{e\}$ if and only if $X^+(G) = \{e\}$. Now assume τ is an isomorphism. Then $X^+(G) = \{e\}$ and, by (4.7), this means $R(G)$ is finitely-generated as a $\mathbb{C}$-algebra. By definition, $\mathscr{G}(G) = G(R(G))$, so $\mathscr{G}(G)$ is an algebraic group with coordinate ring $R(G)$. The category $\mathrm{Mod}(\mathscr{G}(G))$ defined in (2.44) is the category of algebraic $\mathscr{G}(G)$ modules and, by (2.45), $\mathrm{Mod}(G)$ and $\mathrm{Mod}(\mathscr{G}(G))$ are isomorphic categories.

Proposition 4.11 applies in particular to reductive groups G and shows that a reductive group carries a canonical structure of algebraic groups, namely the one given by taking $R(G)$ as coordinate ring.

We have yet to exploit the full strength of (4.7): the algebra D of that theorem has some properties we will further exploit.

Lemma 4.12 Let G be an analytic group, let B be a finitely-generated Hopf subalgebra of $R(G)$ and let T be a torus in $G(B)$ such that $(G(B), e_B, T)$ is a split hull of G. Let $D = \{f \in B \mid t \cdot f = f$ for all $t \in T\}$. Then:

(1) If $f \in D$ and $x \in G$, $f \cdot x \in D$;

(2) D is a finitely-generated $\mathbb{C}$-algebra;

(3) There is a bijection between G and the maximal ideals of D such that the maximal ideal M of G corresponds to the element x in G if $M = \{f \in D \mid f(x) = 0\}$.

Proof. (1) follows from the identity $(t \cdot f) \cdot x = t \cdot (f \cdot x)$ and (2) follows from the isomorphism $D \otimes_{\mathbb{C}} \mathbb{C}[T] \to B$ of (4.5), along with the finite-generation of B. For (3), we first note that if $x \in G$, $M(x) = \{f \in D \mid f(x) = 0\}$ is a maximal ideal of D. In (4.5), we observed that the canonical projection $p: G(B) \to G(B)/T$ induced an injection $p^*: \mathbb{C}[G(B)/T] \to B$ whose image is D. Thus, a maximal ideal M of D corresponds to a unique point y of $G(B)/T$ such that $(p^*)^{-1}M = \{f \in \mathbb{C}[G(B)/T] \mid f(y) = 0\}$. Since $G(B) = e_B(G) \cdot T$ and $e_B(G) \cap T = \{e\}$, there is a unique $x \in G$ with $y = e_B(x)T$. Retracing our identifications shows that $M = \{f \in D \mid f(x) = 0\} = M(x)$.

We can view (4.12) in geometric terms: D can be regarded as the coordinate ring of an affine algebraic variety, and the points of this variety can be identified with G (this is the content of (2) and (3)). Thus, we are regarding the set G as an affine algebraic variety with coordinate ring D. Fix $x \in G$ and consider the function $L_x: G \to G$ given by $L_x(y) = xy$. If $f \in D$, then $(fL_x)(y) = f(xy) = (f \cdot x)(y)$, so $fL_x = f \cdot x$ is in D. This means that $L_x: G \to G$ is a morphism of the affine variety G, and this is true for every x in G (this is the content of (1)). It may be the case that G is not an algebraic group with this variety structure, but we do at least have the left multiplications as morphisms. This geometric interpretation motivates our terminology for the algebras like D.

Definition 4.13 Let G be an analytic group. A *left algebraic group structure* on G is a subalgebra D of $R(G)$ such that:
  (1) if $f \in D$ and $x \in G, f \cdot x \in D$;
  (2) D is a finitely-generated $\mathbb{C}$-algebra;
  (3) there is a bijection between G and the maximal ideals of D such that the maximal ideal M of G corresponds to the element x in G if $M = \{f \in D \mid f(x) = 0\}$.

By (4.12), split hulls yield left algebraic group structures. We will see now that the converse assertion holds.

Proposition 4.14 Let G be an analytic group and suppose D is a left algebraic group structure on G. Let B be the smallest Hopf subalgebra of $R(G)$ containing D. Then there is a torus T in $G(B)$ such that $(G(B), e_B, T)$ is a split hull of G and such that $D = \{f \in B \mid t \cdot f = f$ for all $t \in T\}$.

Proof. We first note that e_B is injective: if $e_B(x) = e$, then $f(x) = f(e)$ for all f in B, so $M(x) = \{f \in D \mid f(x) = 0\} = \{f \in D \mid f(e) = 0\} = M(e)$ and then $x = e$ by (4.13)(3).

Now we note that since D is a right stable subspace of $R(G)$, $\Delta(D) \subseteq R(G) \otimes D$ by (2.21) and, since $\Delta(B) \subseteq B \otimes B$, we actually have $\Delta(D) \subseteq B \otimes D$. We next observe that the inclusion of D into B factors through Δ: if $d \in D$ and $\Delta(d) = \sum b_i \otimes d_i$ where $b_i \in B$ and

$d_i \in D$, then $d(xy) = \sum b_i(x)d_i(y)$ for all $x, y \in G$, so $d(x) = d(xe) = \sum b_i(x)d_i(e)$, or $d = \sum b_id_i(e)$ which means that the composite $D \to B \otimes D \to B$, where the first map is Δ and the second sends $b \otimes d$ to $bd(e)$, is the inclusion.

We regard G as an algebraic variety with coordinate ring D. Then the inclusion $D \subseteq B$ induces a morphism $\phi: G(B) \to G$ of algebraic varieties, such that if $y \in G(B)$, $\phi(y)$ is the unique element of G such that $M(\phi(y)) = \{d \in D \mid \sum b_i(y)d_i(e) = 0$ where $\Delta(d) = \sum b_i \otimes d_i\}$ (here we have used the above factorization of the inclusion through Δ). From this description of ϕ and the properties of Δ, we find that if $x \in G$ then $\phi(e_B(x)) = x$. Also, if $y, z \in G(B)$ and $d \in D$ with $\Delta(d) = \sum b_i \otimes d_i$, then $d(\phi(yz)) = \sum b_i(y)d_i(\phi(z))$. If in this last equation $y = e_B(x)$ then $d(\phi(e_B(x)z)) = \sum b_i(x)d_i(\phi(z)) = d(x\phi(z))$. This holds for all d in D, so we have $\phi(e_B(x)z) = x\phi(z)$. The equation $d(\phi(yz)) = \sum b_i(y)d_i(\phi(z))$ shows that if $\phi(z) = e$, $d(\phi(yz)) = d(\phi(y))$ and again this holds for all d in D, so we have $\phi(yz) = \phi(y)$ if $\phi(z) = e$.

Now let $T = \{z \in G(B) \mid \phi(z) = e\}$. Being a fibre of a morphism of varieties, T is closed in $G(B)$. The formula $\phi(yz) = \phi(y)$ for $y \in G(B)$ and $z \in T$ shows that T is a subgroup of $G(B)$. Since $\phi(e_B(x)) = x$ for all $x \in G$, $e_B(G) \cap T = \{e\}$. Let $y \in G(B)$ and let $x = \phi(y)$. Then $\phi(e_B(x^{-1})y) = x^{-1}\phi(y) = e$ so $e_B(x^{-1})y \in T$ and, hence, $y \in e_B(G)T$. Thus, $G(B)$ is the semidirect product of $e_B(G)$ and T. In particular, $e_B(G)$ is a closed analytic subgroup of $G(B)$ and we have an analytic homomorphism $f: G(B) \to [G(B)/e_B(G)] = T$. By (3.19), there is a torus T' of $G(B)$ such that $G(B) = e_B(G)T'$ and the analytic homomorphism f carries T' onto T. Since T' is a torus and T is an algebraic group, T must also be a torus.

We have now shown that $(G(B), e_B, T)$ is a split hull for G. Let $D' = \{f \in B \mid t \cdot f = f$ for all $t \in T\}$. We need to show that $D' = D$. We recall that $\phi: G(B) \to G$ was induced by the inclusion, which means that if $d \in D$ and $y \in G(B)$, $d(\phi(y)) = d(y)$. Since $\phi(yt) = \phi(y)$ for all $y \in G(B)$, $(t \cdot d)(y) = d(yt) = d(\phi(yt)) = d(\phi(y)) = d(y)$ for all $y \in G(B)$, $t \in T$ and $d \in D$. Thus, D is contained in D'. The inclusions $D \subseteq D' \subseteq B$ correspond to morphisms $G(B) \to G(B)/T \to G$, the composite being ϕ. But $G(B)/T \to G$ is a bijection (the inverse sends x to $e_B(x)T$) which is, in fact, an isomorphism since both varieties are homogeneous and, hence, nonsingular. Thus, the corresponding algebra map $D \subseteq D'$ is also an isomorphism, so $D = D'$.

When we first demonstrated the existence of split hulls in (3.16), we showed that the hulls could be chosen such that there was a reductive complement to a nucleus of the analytic group which commuted with the torus of the split hull. We want to see now how to detect which left algebraic group structures produce, via (4.14), split hulls with the above property.

We need to comment first on representative functions of semi-simple type.

Definition 4.15 Let G be an analytic group and let $f \in R(G)$. Then f is *semisimple* if $[f]_r$ is a completely reducible G-module. $R(G)_S$ is the set of all semisimple f in $R(G)$.

If f is semisimple, let $v \in [f]_r^*$ be evaluation at e, and let $\phi \in ([f]_r^*)^* = [f]_r$ be f. Then $f_{\phi,v}(x) = \phi(xv) = xv(f) = v(x^{-1}(f)) = f(x)$, so f is a representative function associated to a completely reducible G-module. If, conversely, V is a completely reducible G-module, so is V^*, and if we fix $v \in V$, $V^* \to R(G)$ by $\phi \mapsto f_{\phi,v}$ has as its image a right stable subspace of $R(G)$ (which is completely reducible, being an image of V^*), and so for any ϕ in V^*, $[f_{\phi,v}]_r$ is a completely reducible G-module. Thus, the semisimple representative functions are precisely those associated to completely reducible representations.

There is another characterization of semisimple functions which we will need:

Lemma 4.16 Let G be an analytic group, let B be a finitely-generated Hopf subalgebra of $R(G)$ and let U be the unipotent radical of $G(B)$. Then $B \cap R(G)_S = \{f \in B \mid f \cdot u = f \text{ for all } u \text{ in } U\}$.

Proof. Let $f \in B \cap R(G)_S$. Then $[f]_r \subseteq B$, so $[f]_r$ is a $G(B)$-module, and it is easy to check that the action of $G(B)$ on $[f]_r$ is such that $e_B(x)g = g \cdot x^{-1}$ for $g \in [f]_r$ and $x \in G$. Since $e_B(G)$ is dense in $G(B)$ and $[f]_r$ is semisimple as a G-module, $[f]_r$ is semisimple as a

$G(B)$-module. Thus, U acts trivially on $[f]_r$ and, hence, $f \cdot u = f$ for all u in U. Conversely, let $f \in B$ and suppose $f \cdot u = f$ for all u in U. If $x \in G(B)$ and $u \in U$, then $(f \cdot x) \cdot u = f \cdot (xu) = f \cdot (xux^{-1})(x) = f \cdot x$, since $xux^{-1} \in U$. For $x \in G, f \cdot x = f \cdot e_B(x)$, so U acts trivially on $[f]_r$. This means $[f]_r$ is a $G(B)/U$-module and, hence, completely reducible. Every G-submodule of $[f]_r$ is a $G(B)$-module, so $[f]_r$ is a completely reducible G-module.

We retain the notation of (4.16). Let $f \in B \cap R(G)_S$ and let $x \in G$. Then $(x \cdot f) \cdot u = x(f \cdot u) = x \cdot f$ and $(f \cdot x) \cdot u = f \cdot (xux^{-1}) \cdot x = f \cdot x$ for all u in U. Thus, $_l[f]$ and $[f]_r$ are both contained in $B \cap R(G)_S$.

Corollary 4.17 Let G be an analytic group. Then $R(G)_S$ is a Hopf subalgebra of $R(G)$.

Proof. Let $f \in R(G)_S$ and choose a finitely-generated Hopf subalgebra B of $R(G)$ containing f. Since $\Delta(f) \subseteq {}_l[f] \otimes [f]_r$ by (2.4)(d) and, as just remarked, $_l[f]$ and $[f]_r$ are both in $R(G)_S$. Thus, $\Delta(R(G)_S)$ is contained in $R(G)_S \otimes R(G)_S$.

Now assume B in (4.16) actually gives rise to a split hull $(G(B), e_B, T)$ of G. By (4.12), $D = \{f \in B \mid t \cdot f = f$ for all t in $T\}$ is a left algebraic group structure on G. It follows from (4.16) that $D \cap R(G)_S = \{f \in B \mid t \cdot f \cdot u = f$ for all $u \in U$ and $t \in T\}$. We are interested in when $D \cap R(G)_S$ is a Hopf subalgebra of B. By (2.4)(d), this will be the case exactly when $_l[f]_r \subseteq D \cap R(G)_S$ for all f in $D \cap R(G)_S$. Since f in $D \cap R(G)_S$ implies $[f]_r$ is contained in $D \cap R(G)_S$, we need only look for left stability. This property does not always hold; to conveniently refer to it we make the following definition.

Definition 4.18 Let G be an analytic group. A left algebraic group structure D on G is called *basal* if $x \cdot f \in D$ for all $x \in G$ and $f \in R(G)_S \cap D$.

Theorem 4.19 Let G be an analytic group and let B be a finitely-generated Hopf subalgebra of $R(G)$ such that $(G(B), e_B, T)$ is a split hull of G. Then $D = \{f \in B \mid t \cdot f = f$ for all $t \in T\}$ is a basal left algebraic group structure if and only if there is a nucleus K of G and a reductive subgroup P of T such that $G = KP$, $K \cap P = \{e\}$ and $(e_B(P), T) = e$. If there is such a nucleus, it can be taken to be $e_B^{-1}(UT \cap e_B(G))$, where U is the unipotent radical of $G(B)$.

Proof. To simplify notation we identify G and $e_B(G)$ and let H denote $G(B)$, so our split hull is (H, i, T) with i the inclusion. We next assert that D is a basal left algebraic group structure if and only if UT is normal in H: for suppose UT is normal, and let $f \in D' = D \cap R(G)_S = \{f \in B \mid t \cdot f \cdot u = f \text{ for all } t \text{ in } T \text{ and } u \text{ in } U\}$. Then if $x \in G$ and $t \in T$, $t \cdot (x \cdot f) = xt(t^{-1}x^{-1}tx)f = x \cdot t \cdot z \cdot f$, where $z = t^{-1}x^{-1}tx$ is in UT. So $z = us$ with $u \in U$ and $s \in T$, and $t \cdot x \cdot f = x \cdot t \cdot u \cdot s \cdot f = x \cdot t \cdot u \cdot f = x \cdot t \cdot f = x \cdot f$ (the fact that U is normal in H is used to show that $u \cdot f = f$ since $f \cdot u' = f$ for all u' in U). So if UT is normal in H, D' is left stable. Conversely, assume D' is left stable; UT will be normal if $xTx^{-1} \subseteq UT$ for all x in H, so assume there is an x in H with $xtx^{-1} \notin UT$. The variety $H/UT = (H/U)/T$ is affine since H/U is affine and T is reductive and the projection map $p: H \to H/UT$ is a morphism of varieties. We assume $p(xtx^{-1}) \neq p(e)$, so there is $f_0 \in \mathbb{C}[H/UT]$ such that $f_0(p(xtx^{-1})) \neq f_0(p(e))$. Let $f = f_0 p \in B = \mathbb{C}[H]$. Then for $z \in UT$ and $y \in H$, $f(yz) = f(y)$ so $z \cdot f = f$. In particular, if $u \in U$ and $s \in T$, $u \cdot f = f$ (so $f \in R(G)_S$) and $s \cdot f = f$ (so $f \in D$) and, hence, $f \in D'$. Suppose $\Delta(f) = \sum f_i \otimes g_i$ with $f_i \in {}_1[f]$ and $g_i \in [f]_r$. Since D' is left stable, $f_i \in D'$ so $t \cdot f_i = f_i$. Then $f(xtx^{-1}) = \sum (t \cdot f_i)(x)g_i(x^{-1}) = f(xx^{-1}) = f(e)$, contrary to our choice of f_0. Thus, xTx^{-1} is contained in UT and UT is normal.

Next, we show that $UT \cap G$ is always a connected, simply-connected solvable analytic subgroup. Define $\phi: U \to UT \cap G$ as follows: if $u = gt$ with $g \in G$ and $t \in T$ (and every element of H admits a unique expression of this form), then $\phi(u) = ut^{-1}$. It is easy to see that ϕ is a surjection and it is continuous, so $K = UT \cap G$ is a closed connected subgroup of G and it is solvable since UT is. As a subgroup of H, K has a faithful representation and, hence, a nucleus K_0, so $K = K_0 T_0$ where T_0 is a reductive subgroup of K and, hence, a torus. Since $T_0 \subseteq UT$ and T is a maximal torus of UT, there is x in UT with $xT_0x^{-1} \subseteq T$. But $T_0 \subseteq G$ and G is normal in H, so $xT_0x^{-1} \subseteq G$. Thus, $xT_0x^{-1} \subseteq T \cap G = \{e\}$ so $T_0 = e$ and $K = K_0$ is simply-connected.

Now if UT is normal in H, K is normal in G and we have an injection $\psi: G/K \to H/UT$. But H/UT is reductive and, since $H = GT$, ψ is onto, so G/K is reductive. Thus, K is a nucleus of G. There is, thus, a reductive subgroup P of G with $G = KP$ and $P \cap K = \{e\}$. Then $P \cap UT = (G \cap UT) \cap P = K \cap P = \{e\}$, so $P \to H/UT$ by ψ is an isomorphism. Now P is contained in a maximal reductive subgroup Q of H and there is $x \in H$ such that $T \subseteq xQx^{-1}$. Since both K and G are normal in H, we can replace P by xPx^{-1}, so there is a maximal reductive subgroup Q of H containing both P and T. Then $(P, T) \subseteq Q$, but also $(P, T) \subseteq U$ since T is normal, and, hence, central, in H/U. Thus, $(P, T) \subseteq U \cap Q = \{e\}$.

Conversely, assume there is a nucleus K of G and a reductive subgroup P of G with $(P, T) = \{e\}$. Then TP is a reductive subgroup of H and, hence, is contained in a maximal reductive subgroup Q. Now K is normal in H and $TP \to H/K$ is an isomorphism, so H/K is reductive, hence, algebraic. Thus the analytic homomorphism $Q \to H/K$ is also algebraic, so its kernel $K \cap Q$ is an algebraic subgroup of Q. Now $K \cap Q$ is normal in Q, hence reductive, and solvable, so its identity component is a torus. Since K has no elements of finite order, this torus is trivial so $K \cap Q$ is finite and, hence, trivial. We have $K \cap Q = \{e\}$, $TP \subseteq Q$ and $H = KTP$, so $TP = Q$. Since also $H = UQ$, $H = UTP$. Now UT normalizes UT and $(P, UT) \subseteq U$ since $(P, T) = e$, so UT is normal in H, as we wanted to show.

There are a couple of observations worth making about (4.19) and its proof. First, if the split hull $(G(B), e_B, T)$ yields a basal left algebraic group structure, then there is a determined nucleus of G, namely $e_B^{-1}(UT \cap e_B(G))$. Now, in general, analytic groups that have nuclei have many of them and they bear no particular relation to a split hull. But now we know that a split hull of a certain type does yield a canonical nucleus. In (3.16), we showed that if G is an analytic group with nucleus K and H a reductive subgroup of G with $G = KH$ and $K \cap H = \{e\}$, then G has a split hull $(\bar{G}, f, T)$ with $(T, f(H)) = \{e\}$. Thus, by (4.19), $K = f^{-1}(UT \cap f(G))$ where

U is the unipotent radical of $\bar{G}$ so, in fact, every nucleus of G arises by the procedure of (4.19).

Next, we also want to note that the criterion developed in the proof of (4.19) for the existence of the reductive subgroup P of G complementary to the nucleus and commuting with T depends only on the split hull $(G(B), e_B, T)$, i.e., that UT is normal in $G(B)$. We now use this to give an example of an analytic group and a split hull such that there is no reductive subgroup of the desired type.

Example. Let Γ be the subgroup $\{(1, I), (-1, -I)\}$ of $\mathbb{C}^* \times SL_2\mathbb{C}$, let $Q = (\mathbb{C}^* \times SL_2\mathbb{C})/\Gamma$ and let $p: \mathbb{C}^* \times SL_2\mathbb{C} \to Q$ be the projection; Q is a reductive algebraic group. Let $G = \mathbb{C} \times SL_2\mathbb{C}$, let $H = \mathbb{C} \times Q$ and define $f: G \to H$ by $f(t, A) = (t, p(e^t, A))$. It is easy to see that f is injective and that the projection of $f(G)$ onto each factor $\mathbb{C}$ and Q of H is onto. Let $\overline{f(G)}$ be the Zariski-closure of $f(G)$. Then $\overline{f(G)}$ projects onto both $\mathbb{C}$ and Q; since $\mathbb{C}$ is the unipotent radical of H and Q is the unique maximal reductive subgroup of H, we conclude $\overline{f(G)} = H$. Define $h: \mathbb{C}^* \to Q$ by $h(t) = p(t, \mathrm{diag}(t^2, t^{-2}))$; h is injective (and algebraic), so its image is a torus in Q. Let $T = \{0\} \times h(\mathbb{C}^*)$ in H. Then $T \cap f(G) = e$, while $f(G)T = H$. The first equality is left to the reader. For the second, note that $(t, p(s, A)) = f(t, \mathrm{diag}(e^{-t}s, e^t s^{-1})^{-2} A)(0, h(e^{-t}s))$. Thus, (H, f, T) is a split hull of G. The unipotent radical U of H is $\mathbb{C} \times 0$ and UT is not normal in H. Thus, we conclude, by (4.19), that the left algebraic group structure D on G arising, via (4.12), from the split hull (H, f, T) is not basal.

The group G of the above example was even algebraic. To get a nonbasal left algebraic group structure, we clearly need a nonreductive group, so we need at least a one-dimensional nucleus. We will see next that we also need a nonsolvable reductive part, so that the above example is truly minimal.

Proposition 4.20 Let G be a solvable analytic group. Then every left algebraic group structure on G is basal.

Proof. Let (H, f, T) be a split hull of G and let U be the unipotent radical of H; T is contained in a maximal torus S of H and $H = US$. Both U and S normalize UT, so UT is normal in H. Now (4.14) and (4.19) imply the result.

The reader may be wondering about the effort being expended on basal left algebraic group structures. As we will see in Chapter 5, these are the structures that admit convenient linearization in terms of Lie algebras. Our final result in this chapter shows that we do not lose very much by restricting our attention to these structures.

Proposition 4.21 Let G be an analytic group with a faithful representation and let F be a finite subset of $R(G)$. Then there is a finitely-generated Hopf subalgebra B of $R(G)$ containing F and a torus T in $G(B)$ such that $(G(B), e_B, T)$ is a split hull of G where the left algebraic group structure $D = \{f \in B \mid t \cdot f = f \text{ for all } t \in T\}$ is basal.

Proof. Suppose (H_1, f_1, T_1) and (H_2, f_2, T_2) are split hulls of G and $p : H_1 \to H_2$ is an algebraic group homomorphism such that $pf_1 = f_2$ and $T_1 = p^{-1}(T_2)$. Let U_i be the unipotent radical of H_i. Then p is surjective, so $p(U_1) = U_2$ and, since $\operatorname{Ker}(p) \subseteq T_1$, $p^{-1}(U_2 T_2) = U_1 T_1$. So $U_2 T_2$ normal in H_2 implies $U_1 T_1$ normal in H_1. Thus, if the left algebraic group structure associated to (H_2, f_2, T_2) is basal, so is the one associated to (H_1, f_1, T_1). By (3.16), there is a split hull of G with associated left algebraic group structure basal (we are using (4.19) to intepret (3.16) here). So there is a finitely-generated Hopf subalgebra B_0 of $R(G)$ and a torus T_0 in $G(B_0)$ such that $(G(B_0), e_{B_0}, T_0)$ is a split hull of G where the left algebraic group structure $D_0 = \{f \in B_0 \mid t \cdot f = f \text{ for all } t \in T_0\}$ is basal. Now let B be a finitely-generated Hopf subalgebra of $R(G)$ containing B_0 and F, and let T be the inverse image of T_0 in $G(B)$, where $G(B) \to G(B_0)$ is the surjection induced by the inclusion $B_0 \subseteq B$. Then $(G(B), e_B, T)$ is a split hull of G (3.17) and the associated left algebraic group structure D is basal.

Summary of Results of Chapter 4

The algebra of representative functions on an analytic group is Hopf-algebra isomorphic to the algebra of representative functions on a group with a faithful representation, so for the structure theory of the algebra we may assume faithful representations. If G is an analytic group and $(\bar{G}, f, T)$ is a split hull of G, then $\mathbb{C}[\bar{G}]$ is, as an algebra, isomorphic to the group ring $\mathbb{C}[\bar{G}/T][X(T)]$. By passing to direct limits, we then have that if G is an analytic group with a faithful representation, $R(G)$, as an algebra, is isomorphic to the group ring $D[Q]$, where Q is the group of characters of G factoring as $G \to \mathbb{C} \to \mathbb{C}^*$ and D is a finitely-generated $\mathbb{C}$-subalgebra of $R(G)$ which is right stable, and such that there is a bijection between G and the maximal ideals of D, where the maximal ideal M corresponds to the element x in G if $M = \{f \in D \,|\, f(x) = 0\}$.

Paralleling this description of $R(G)$ as an algebra is a similar decomposition of $\mathscr{G}(G)$ as a group: $\mathscr{G}(G) = \tau(G) \cdot T$ where T is a subgroup of $\mathscr{G}(G)$ with $T \cap \tau(G) = \{e\}$ and, as an abstract group, T is isomorphic to $\mathrm{Hom}(X^+(G), \mathbb{C}^*)$.

Algebras D in the above description are called *left algebraic group structures* on G. They arise from, and give rise to, split hulls.

A function f in $R(G)$ is *semisimple* if $[f]_r$ is a completely reducible G-module; $R(G)_S$ is the set of all semisimple representative functions on G. A left algebraic group structure D is *basal* if $D \cap R(G)_S$ is left stable. A left algebraic group structure is basal if and only if it arises from a split hull $(\bar{G}, f, T)$ such that there is a nucleus K of G and a reductive subgroup H of G with $KH = G$, $K \cap H = \{e\}$ and $(T, f(H)) = \{e\}$.

5
Left algebraic groups

In Chapter 4, we saw that the algebra $R(G)$ of representative functions on the analytic group G could be decomposed as a group ring of the abelian group $Q = \exp(\text{Hom}(G, \mathbb{C}))$ with coefficients from any left algebraic group structure A in $R(G)$. We saw further that left algebraic group structures arise from and give rise to split hulls of G. Thus, the study of the possible decompositions $R(G) = A[Q]$ is equivalent to the search for the split hulls of G. (We will restrict our attention to basal left algebraic group structures and their corresponding split hulls.) In Chapter 3, we saw how to construct a split hull of an analytic group; a review of that construction will reveal that the only choices made were the selection of a nucleus and essentially the choice of a Cartan subgroup of that nucleus (a Cartan subgroup is an analytic subgroup whose Lie algebra is a Cartan subalgebra). Moreover, that construction meets our criterion (4.19) so that the corresponding left algebraic group structure is basal. This suggests that if we want to reverse our construction, we begin with a basal left algebraic group structure, then pass to the associated split hull, and then we must produce a nucleus and a Cartan subgroup of it. That program turns out to be possible, and we carry it out in this chapter.

There are some additional consequences: the identification of nuclei and their Cartan subgroups can be done strictly in Lie algebra terms, once the Lie algebra of a maximal reductive subgroup is designated. This means that the decompositions $R(G) = A[Q]$ can, at least in principle, be obtained already from $\text{Lie}(G)$.

We begin by sharpening the correspondences (4.12) and (4.14) between left algebraic group structures and split hulls.

Lemma 5.1 Let G be an analytic group, let B be a finitely-generated Hopf subalgebra of $R(G)$ and suppose T is a torus in

$G(B)$ such that $(G(B), e_B, T)$ is a split hull of G. Let $D = \{f \in B \mid t \cdot f = f$ for all $t \in T\}$. Then $T = \{x \in G(B) \mid x \cdot f = f$ for all $f \in D\}$.

Proof. Let $d : G(B) \to T$ be the algebraic group homomorphism of (4.3), and let H be the kernel of d. Assume $x \in G(B)$ is such that $x \cdot f = f$ for all $f \in D$. Then $x = ht$ for some $h \in H$ and $t \in T$, and to show that $x \in T$ we may assume $x = h$ is in H. In (4.5) we saw that $D \otimes \mathbb{C}[T] \to B$ by $a \otimes f \mapsto a(fd)$ is a $\mathbb{C}$-algebra isomorphism. If $a \in D$, $a(x) = (x \cdot a)(e) = a(e)$, and if $f \in \mathbb{C}[T]$, $(fd)(x) = f(e) = (fd)(e)$, since $x \in H$. It follows that $p(x) = p(e)$ for all $p \in B$, so $x = e$. Thus, $\{x \in G(B) \mid x \cdot f = f$ for all $f \in D\}$ is contained in T and, hence, equals T.

Definition 5.2 Let G be an analytic group. A split hull $(\bar{G}, f, T)$ of G is *reduced* if T contains no central elements of $\bar{G}$ except e.

Lemma 5.3 Let G be an analytic group, let B be a finitely-generated Hopf subalgebra of $R(G)$ and suppose T is a torus in $G(B)$ such that $(G(B), e_B, T)$ is a split hull of G. Let $D = \{f \in B \mid t \cdot f = f$ for all $t \in T\}$. Then B is the smallest Hopf subalgebra of $R(G)$ containing D if and only if $(G(B), e_B, T)$ is reduced.

Proof. Let B_0 be a Hopf subalgebra of B containing D, and let $p : G(B) \to G(B_0)$ be the algebraic group epimorphism induced from the inclusion $B_0 \subseteq B$. Let $T_0 = \{x \in G(B_0) \mid x \cdot f = f$ for all $f \in D\}$. It follows from (5.1) that $p^{-1}(T_0) = T$. In particular, the kernel L of p is contained in T. Since $G(B) = e_B(G)T$, $G(B_0) = G(B)/L = e_B(G) \cdot T/L = e_{B_0}(G) \cdot T_0$ and $T_0 \cap e_{B_0}(G) = \{e\}$. If $x \in G$ and $y \in L$, $z = ye_B(x)y^{-1}$ is in $e_B(G)$ and $p(z) = e_{B_0}(x)$, so $z = e_B(x)$. Thus, L centralizes $e_B(G)$ and, since $e_B(G)$ is Zariski-dense in $G(B)$, it follows that L is central in $G(B)$. So if $B_0 \neq B$, $(G(B), e_B, T)$ is not reduced. Conversely, if $(G(B), e_B, T)$ is not reduced, and N is the intersection of T and the center of $G(B)$, let $B_0 = B^N$. Then B_0 is a Hopf subalgebra of $R(G)$ with $D \subseteq B_0 \subseteq B$ and since $N \neq \{e\}$, $B_0 \neq B$. So B is not the smallest Hopf subalgebra of $R(G)$ containing D.

Lemmas (5.2) and (5.3) combine to yield a one–one correspondence between reduced split hulls and left algebraic group structures:

Theorem 5.4 Let G be an analytic group. Then there is a one–one correspondence between left algebraic group structures on G and reduced split hulls of G, where the left algebraic group structure D corresponds to the reduced split hull $(G(B), e_B, T)$, with B the smallest Hopf subalgebra of $R(G)$ containing D and $T = \{x \in G(B) | x \cdot f = f$ for all $f \in D\}$. Moreover, D is basal if and only if UT is normal in $G(B)$, where U is the unipotent radical of $G(B)$.

Proof. If we begin with D, then $(G(B), e_B, T)$ is a split hull by (4.14), and T is as described by (5.1). By (5.3), this split hull is reduced. Conversely, (5.3) and (4.12) show that if we start with a reduced split hull the D we produced yields B back. The final assertion, about basal subalgebras, comes from (4.19).

We intend to linearize basal left algebraic group structures by studying reduced split hulls. A split hull of G is a triple $(\bar{G}, f, T)$ and the property of being reduced was defined in terms of the designated torus T. The following example shows that being reduced really does depend on the designated torus, and not on the algebraic group $\bar{G}$ alone, or even on the pair $(\bar{G}, f)$.

Example. Let $\bar{G}$ be the algebraic group of quadruples (a, b, s, t) where $a, b \in \mathbb{C}$, $s, t \in \mathbb{C}^*$ and $(a, b, s, t)(c, d, v, w) = (a + stc, b + d, sv, tw)$. Let G denote the analytic subgroup of G of quadruples (a, b, e^b, t) where $a, b \in \mathbb{C}$ and $t \in \mathbb{C}^*$. Let $U = \{(a, b, 1, 1) | a, b \in \mathbb{C}\}$, let $Q = \{(0, 0, s, t) | s, t \in \mathbb{C}^*\}$, let $P = \{(0, 0, 1, t) | t \in \mathbb{C}^*\}$, let $T = \{(0, 0, s^2, s^{-1}) | s \in \mathbb{C}^*\}$ and let $T' = \{(0, 0, s^{-1}, s) | s \in \mathbb{C}^*\}$. Then U is the unipotent radical of $\bar{G}$, G is Zariski-dense in $\bar{G}$, and if i is the inclusion, $(\bar{G}, i, T)$ and $(\bar{G}, i, T')$ are split hulls of G; P is a maximal reductive subgroup of G and $(P, T) = (P, T') = \{e\}$; Q is a maximal reductive subgroup of $\bar{G}$ with $\bar{G} = UQ$, and T, T' are contained in Q, so an element of Q is central if and only if it centralizes U. It is easy to compute that $(0, 0, s, t) \times$

$(a, b, 1, 1)(0, 0, s^{-1}, t^{-1}) = (sta, b, 1, 1)$. Thus, $(0, 0, s, t)$ is central if and only if $st = 1$. Thus no nonidentity elements of T are central in $\bar{G}$, so $(\bar{G}, i, T)$ is a reduced split hull of G, while every element of T' is central, so $(\bar{G}, i, T')$ is not reduced. We can, of course, assume that $\bar{G} = G(B)$ for a suitable Hopf subalgebra B of $R(G)$, and that $i = e_B$. Then $B^T = D$ and $B^{T'} = D'$ are basal left algebraic group structures on G contained in B. By (5.3), B is the smallest Hopf subalgebra containing D, but there is a proper Hopf subalgebra of B containing D': in fact, it is D' itself. For D' is the coordinate ring of the algebraic group $\bar{G}/T'$, to which G is isomorphic via the inclusion. ($\bar{G}/T'$ is the algebraic group of triples (a, b, x) with a, $b \in \mathbb{C}$ and $x \in \mathbb{C}^*$ with $(a, b, x)(c, d, y) = (a + xc, b + d, xy)$; the map $\bar{G} \to \bar{G}/T'$ sends (a, b, s, t) to (a, b, st).)

As the example makes clear, in the correspondence between left algebraic group structures and reduced split hulls we must keep track of the designated tori. It is convenient to do so via the corresponding nuclei:

Proposition 5.5 Let G be an analytic group and let D be a basal left algebraic group structure on G. Let B be the smallest Hopf subalgebra of $R(G)$ containing D, let U be the unipotent radical of $G(B)$, let $T = \{x \in G(B) \mid x \cdot f = f$ for all $f \in D\}$ and let $K = e_B^{-1}(UT)$. Then K is a nucleus of G and (UT, e_B, T) is a reduced split hull of K.

Proof. We identify G with its image in $G(B)$ so e_B is inclusion. By (4.19), K is a nucleus of G. Let P be a reductive subgroup of G with $KP = G$, $K \cap P = \{e\}$, and $(P, T) = \{e\}$. Let $\bar{K}$ be the Zariski-closure of K. Then $\bar{K} \subseteq UT$ and $\bar{K} \cdot P$ is Zariski-closed and contains G, so $\bar{K} \cdot P = \bar{G} = (UT) \cdot P$. So $\bar{K} = UT$. Also, $\bar{G} = G \cdot T = KT \cdot P \subseteq UT \cdot P = \bar{G}$, so $KT = UT$. Thus, (UT, e_B, T) is a split hull of K. If $x \in T$ is central in UT, x is central in $\bar{G}$ since $(P, T) = \{e\}$, so $x = e$ since $(G(B), e_B, T)$ is reduced by (5.2).

Now because of (5.5), we will look at reduced split hulls of simply-connected solvable analytic groups. The theory here is contained in the next theorem.

Theorem 5.6 Let K be a simply-connected solvable analytic group and let (G, f, T) be a split hull of K. Let $C = \{x \in G \mid xt = tx$ for all $t \in T\}$. Then:

(1) $f(K) \cap C$ is a Zariski-dense nucleus of C;

(2) $f(K) \cap C$ determines T;

(3) $f(K) \cap C$ is the unique connected analytic subgroup K_0 of C such that:
   (i) K_0 is contained in C;
   (ii) Lie $(f^{-1}(K_0))$ is a Cartan subalgebra of Lie(K).

Proof. We assume that f is inclusion and that K is a subgroup of G. Then, since $G = K \cdot T$, we have $C = (K \cap C) \cdot T$. Moreover, C is a Cartan subgroup of the algebraic group G, so C maps surjectively to G^{ab}. Now, if U is the unipotent radical of G, then C/T maps onto $U/(G, G)$. Since $(G, G) = (K, K)$, we conclude that $K \cap C$ maps onto $U/(K, K)$. Since C is connected, so is $K \cap C$, and we have $\dim(K \cap C) \geq \dim(U/(K, K)) = \dim(U) - \dim((K, K)) = \dim(K) - \dim((K, K)) = \dim(K^{ab})$. Let $K' = (K \cap C)(K, K)$. Then K' is a connected analytic subgroup of K and $\dim(K'/(K, K)) \geq \dim(K^{ab})$ so $K' = K$. Let $\overline{(K \cap C)}$ be the Zariski-closure of $K \cap C$ in G. Then $(K, K)\overline{(K \cap C)}$ is Zariski-closed and contains K, so $(K, K)\overline{(K \cap C)} = G$. Moreover, $\overline{(K \cap C)}$ is contained in $C = (K \cap C) \cdot T$, so $\overline{(K \cap C)} = (K \cap C) \cdot T_1$ where $T_1 \subseteq T$ is a torus. Now $G = K \cdot T$ and $G = (K, K)\overline{(K \cap C)} = (K, K)(K \cap C) \cdot T_1 = K \cdot T_1$ so $T_1 = T$. We have now shown that $K \cap C$ is a Zariski-dense, solvable normal analytic subgroup of C, and $C/K \cap C = T$ is a torus, so $K \cap C$ is a nucleus of C and (1) follows. Now T is the unique maximal torus of C, so since $K \cap C$ is Zariski-dense in C, $K \cap C$ determines T.

Next, since C is a nilpotent group we have that $K \cap C$ is a nilpotent group. The difficult point in the proof of (3) is in showing that $K \cap C$ satisfies (ii). We assume for the moment that Lie$(C \cap K)$ is a Cartan subalgebra of Lie(K), and that K_0 is a connected analytic subgroup of C such that K_0 is contained in C and Lie(K_0) is a Cartan subalgebra of Lie(K). Then K_0 is contained in $K \cap C$, so Lie(K_0) is contained in Lie$(K \cap C)$, and, thus, they are equal since $\text{Lie}(K_0)$ is a Cartan subalgebra (hence, maximal nilpotent) in $\text{Lie}(K)$, so $K_0 = K \cap C$.

Now we show that $\text{Lie}(C \cap K)$ is Cartan in $\text{Lie}(K)$. Since K is a normal subgroup of G, $\text{Lie}(K)$ is a G-subspace of $\text{Lie}(G)$ under the adjoint action Ad. Since the adjoint action is algebraic, and $K \cap C$ is Zariski-dense in C, any $\text{Ad}(K \cap C)$-subspace of $\text{Lie}(K)$ is $\text{Ad}(C)$ stable, and if $K \cap C$ acts unipotently on a stable subspace so does C. It follows that $\{v \in \text{Lie}(K) \mid \text{ad}(X)^n v = 0 \text{ for all } X \in \text{Lie}(C)$ (and some $n)\} = \{v \in \text{Lie}(K) \mid \text{ad}(X)^n v = 0$ for all $X \in \text{Lie}(C \cap K)$ (and some $n)\}$. Call this space $\text{Lie}(K)_0$. Since C is a Cartan subgroup of G, $\text{Lie}(C)$ is a Cartan subalgebra of $\text{Lie}(G)$, so $\text{Lie}(C) = \{v \in \text{Lie}(G) \mid \text{ad}(X)^n v = 0$ for all $X \in \text{Lie}(C)\}$. Since $\text{Lie}(G) = \text{Lie}(K) + \text{Lie}(T)$ and T and C commute, it follows that $\text{Lie}(C) = \text{Lie}(K)_0 + \text{Lie}(T)$. Thus, $\text{Lie}(C) \cap \text{Lie}(K) = \text{Lie}(K)_0$, or $\text{Lie}(C \cap K) = \text{Lie}(K)_0$. This shows that $\text{Lie}(C \cap K)$ is a Cartan subalgebra of $\text{Lie}(K)$.

Definition 5.7 Let G be an analytic group. A *Cartan subgroup* C of G is an analytic subgroup such that $\text{Lie}(C)$ is a Cartan subalgebra of $\text{Lie}(G)$.

In (5.6), we began with a split hull (G, f, T) of the simply-connected solvable analytic group K and obtained a (unique) Cartan subgroup E of K which enabled us to recover T. If $G = G(B)$ and $f = e_B$ for a finitely-generated Hopf subalgebra B of $R(G)$, then $D = \{f \in B \mid t \cdot f = f \text{ for all } t \in T\}$ is the corresponding left algebraic group structure on K. We would like to describe D directly in terms of E; this will give a direct construction of left algebraic group structures from Cartan subgroups of K. In fact, this even gives left algebraic group structures from Cartan subalgebras of $\text{Lie}(K)$. This direct description will be a consequence of the following lemma.

Lemma 5.8 Let K be a simply-connected solvable analytic group and let (G, f, T) be a split hull of K. Let $C = \{x \in G \mid xt = tx \text{ for all } t \in T\}$, and let $E = f^{-1}(f(K) \cap C)$. Let V be a finite-dimensional complex vector space, let $p: G \to \text{GL}(V)$ be an algebraic homomorphism and let $v \in V$. Then $p(t)(v) = v$ for all t in T if and only if $pf(E)$ is unipotent on $W = \langle p(c)(v) | c \in C \rangle$.

Proof. We regard W as an algebraic C-module via p and regard f as inclusion. Then, $C = E \times T$, and E is Zariski-dense in C by (5.6)(1). Let U be the unipotent radical of C, so $C = U \times T$. If $tv = v$ for all $t \in T$, then, since T is central in C, T acts trivially on all of W, so under the canonical (algebraic) homomorphism $q: C \to \mathrm{GL}(W)$ we have $q(C) = q(U)$ and C acts unipotently on W and, hence, E acts unipotenily on W. If conversely, E acts unipotently on W, $q(E)$ is contained in a unipotent algebraic subgroup H of $\mathrm{GL}(W)$. Since E is Zariski-dense in C, $q(E)$ is Zariski-dense in $q(C)$ so $q(C)$ is contained in H. Then $q(T)$ must be trivial, so $tv = v$ for all t in T.

We will be interested in applying (5.8) when V is a subspace of $R(G)$, so we make the following definition.

Definition 5.9   Let G be an analytic group, let H be a subgroup and let f be in $R(G)$. We say that H *acts unipotently on* f if on the H-module $V = \langle x \cdot f | x \in H \rangle$ the linear maps $v \mapsto x \cdot v$ are unipotent for all x in H. We let $R_u[G; H]$ be the set of all f in $R(G)$ on which H acts unipotently.

Theorem 5.10   Let K be a simply-connected solvable analytic group and let B be a finitely-generated Hopf subalgebra of $R(G)$ and T a torus in $G(B)$ such that $(G(B), e_B, T)$ is a split hull of K. Let $C = \{x \in G(B) | xt = tx \text{ for all } t \text{ in } T\}$, let $E = e_B^{-1}(e_B(K) \cap C)$ and let $D = \{f \in B | t \cdot f = f \text{ for all } t \text{ in } T\}$. Then, $D = R_u[K; E]$.

Proof. Let $S = \{f \in B | E \text{ acts unipotently on } f\}$. We show first that $S = D$. Let $f \in B$ and let $W = \langle x \cdot f | x \in E \rangle$. Since $e_B(E)$ is Zariski-dense in C by (5.6)(1), it follows that $W = \langle x \cdot f | x \in C \rangle$. Now by (5.8) we have that $f \in D$ if and only if $f \in S$ (regarding W as a submodule of the algebraic $G(B)$-module $[f]$). Now let B' be a finitely-generated Hopf-subalgebra of $R(G)$ containing B, let $p: G(B') \to G(B)$ be the algebraic group homomorphism induced by the inclusion $B \subseteq B'$ and let $T' = p^{-1}(T)$. By (4.4), $(G(B'), e_{B'}, T')$ is a split hull of G also. Let $C' = \{x \in G(B') | xt = tx$

for all t in T'}, let $E' = e_{B'}^{-1}(e_{B'}(K) \cap C')$, let $D' = \{f \in B' | t \cdot f = f$ for all t in $T'\}$ and let $S' = \{f \in B' | E'$ acts unipotently on $f\}$. It follows that $D' = S'$. By (4.4)(i), $D' = D$. We are going to see that $S' = S$ also, so $D = S'$.

Since $p(T') = T$, we have $p(C') \subseteq C$, so $p(e_{B'}(K) \cap C') \subseteq e_B(K) \cap C$. It follows that $E' \subseteq E$. But by (5.6), both E' and E are Cartan subgroups of K, so $E' = E$ and, hence, $S' = S$. Now $R(K) =$ dir lim$\{B' | B'$ is a finitely-generated Hopf subalgebra of $R(K)$ containing $B\}$, so we indeed have that $D = R_u[K; E]$.

We note that in (5.10) the description of the left algebraic group structure D is given directly in terms of K and its associated Cartan subgroup E: the split hull $(G(B), e_B, T)$ does not intervene. We next will see that every Cartan subgroup of K yields a left algebraic group structure.

Lemma 5.11 Let K be a simply-connected solvable analytic group, let E be a Cartan subgroup of K and let $y \in K$. Then $y \cdot R_u[K; E] = R_u[K; yEy^{-1}]$.

Proof. Let $f \in R(K)$, let $V = \langle x \cdot f | x \in E \rangle$, let $E' = yEy^{-1}$ and let $V' = \langle x \cdot y \cdot f | x \in E' \rangle$. For $v \in V$, let $t(v) = y \cdot v$. Since $y \cdot x \cdot f = (yxy^{-1}) \cdot y \cdot f$, $t(V) = V'$. Define $s : E \to E'$ by $s(x) = yxy^{-1}$. Then for $v \in V$ and $x \in E$, $t(xv) = s(x)t(v)$ (since $y \cdot xv = yxy^{-1} \cdot yv$), so E acts unipotently on V if and only if E' acts unipotently on V'. Thus, f is in $R_u[K; E]$ if and only if $y \cdot f$ is in $R_u[K; yEy^{-1}]$.

Lemma 5.12 Let G be an analytic group, let D be a left algebraic group structure on G and let $x \in G$. Then $x \cdot D$ is also a left algebraic group structure on G.

Proof. We will verify the three conditions of Definition (4.13) for $x \cdot D$. First, if $y \in G$, we have $D \cdot y \subseteq D$, so $(x \cdot D) \cdot y = x \cdot (D \cdot y)$ is contained in $x \cdot D$, so $x \cdot D$ satisfies (4.13)(1). Since $x \cdot D$ is isomorphic to D as a $\mathbb{C}$-algebra, and D is finitely-generated as $\mathbb{C}$-algebra, so is $x \cdot D$ and (4.13)(2) follows. Now suppose M is a maximal ideal of $x \cdot D$. Since $x \cdot D$ is finitely-generated as a $\mathbb{C}$-algebra, $(x \cdot D)/M = \mathbb{C}$, and we let $q : D \to \mathbb{C}$ be the composite $f \mapsto x \cdot f \mapsto x \cdot f + M$. Then q is a $\mathbb{C}$-algebra homomorphism, so there is a unique element y of G such that $q(f) = f(g)$ for all f in D. Now $x \cdot f \in M$ if and only if $q(f) = 0$, and $q(f) = 0$ if and only if $f(y) = 0$, so $M = \{h \in x \cdot D \,|\, h(yx^{-1}) = 0\}$, and (4.13)(3) holds for $x \cdot D$ also.

Corollary 5.13 Let K be a simply-connected solvable analytic group and let E be a Cartan subgroup of K. Then $R_u[K; E]$ is a left algebraic group structure on K.

Proof. K has a nucleus (itself) so by (3.16), K has a split hull and, hence, K satisfies the hypotheses of (5.10). From (5.10) and (5.6), we have that there is a left algebraic group structure D on K where $D = R_u[K; E']$ for some Cartan subgroup E' of K. Since Cartan subgroups are conjugate, there is x in K with $E = xE'x^{-1}$. By (5.11), $x \cdot D = R_u[K; E]$, and then by (5.12) $x \cdot D$ is also a left algebraic group structure on K.

We can actually extract a bit more information out of the above results, as we shall see in (5.16) below. First a comment on (5.13): in the notation of that result, given E, we have the left algebraic group structure $R_u[K; E]$, which leads by (4.14) to a reduced split hull. By (5.6), this reduced split hull yields a Cartan subgroup E' of K, and by (5.10) we have $R_u[K; E] = R_u[K; E']$. We want to see the relation between E and E'; they turn out to be equal, as we will now see.

Lemma 5.14 Let G be an algebraic group with coordinate ring B, let T be a torus in T, let $D = \{f \in B \,|\, t \cdot f = f$ for all t in $T\}$, let N be the normalizer of T in G and let $M = \{x \in G \,|\, x \cdot D \subseteq D\}$. Then $N = M$.

Proof. Clearly N is contained in M. Suppose y in G is not in N. Then $yT \neq Ty$, so there is $t_0 \in T$ such that $t_0 y \notin yT$ and, hence, the cosets $t_0 yT$ and yT are disjoint. Now D can be regarded as the coordinate ring of the affine variety G/T, so there must be $f \in D$ with $f(t_0 y) \neq f(y)$. If we had $y \in M$, then $y \cdot f = t_0 \cdot y \cdot f$, so $f(y) = (y \cdot f)(e) = (t_0 \cdot y \cdot f)(e) = f(t_0 y)$, contrary to our choice of f. Thus, $y \notin M$. It follows that $N = M$.

Proposition 5.15 Let K be a simply-connected solvable analytic group and let E be a Cartan subgroup of K. Then $E = \{x \in K \mid x \cdot R_u[K; E] \subseteq R_u[K; E]\}$.

Proof. By (5.13), $D = R_u[K; E]$ is a left algebraic group structure on K. Let B be the smallest Hopf subalgebra of $R(K)$ containing D. Let $G = G(B)$. By (4.14) there is a torus T in G with (G, e_B, T) a split hull of K and $D = \{f \in B \mid t \cdot f = f \text{ for all } t \in T\}$. By (5.14), $M = \{x \in G \mid x \cdot D \subseteq D\}$ is the normalizer of T in G. But G is connected solvable, so M is the centralizer of T in G. Thus, $E' = e_B^{-1}(e_B(K) \cap M)$ is a Cartan subgroup of K by (5.6), and $E' = \{x \in K \mid x \cdot R_u[K; E] \subseteq R_u[K; E]\}$. If $x \in E$, we have then that $x \in E'$ by (5.11), since then $xEx^{-1} = E$. Thus, $E \subseteq E'$. Since both are Cartan subgroups of K, $E = E'$ and (5.15) follows.

We can now summarize in full the description of left algebraic group structures in the simply-connected solvable case.

Theorem 5.16 Let K be a simply-connected solvable analytic group.
   (i) There is a one–one correspondence between Cartan subgroups of K and left algebraic group structures on K such that the Cartan subgroup E of K corresponds to the left algebraic group structure $R_u[K; E]$ and the left algebraic group structure D corresponds to the Cartan subgroup $\{x \in K \mid x \cdot D \subseteq D\}$.
   (ii) If D and D' are left algebraic group structures on K there is $x \in K$ with $x \cdot D = D'$.
   (iii) There is a unique finitely-generated Hopf subalgebra B of $R(G)$ such that $G(B)$ contains a torus T with $(G(B), e_B, T)$ a reduced split hull of K.

Proof. (i) follows from (5.10), (5.13) and (5.15).
   (ii) follows from (i), (5.11) and the conjugacy of Cartan subgroups.
   (iii) follows from (ii) and (5.3): the smallest Hopf subalgebra containing any left algebraic group structure must contain them all by (ii).

It is tempting to refer to the split hull of (iii) as the unique reduced split hull of K. This is not quite accurate: the choice of the torus T is not unique. But the idea is correct: the pair $(G(B), e_B)$ is unique. This means, for example, that if $a : K \to K$ is an analytic automorphism, there is a unique algebraic group automorphism $a' : G(B) \to G(B)$ with $e_B a = a' e_B$.

In case K is not only solvable, but actually nilpotent, then K is a Cartan subgroup of itself and (5.16) takes the following form:

Corollary 5.17 Let N be a simply-connected nilpotent analytic group. Then there is a unique left algebraic group structure D on N, and D is a Hopf subalgebra of $R(N)$.

Proof. N has at least one left algebraic group structure D. By (5.16)(i), $E = \{x \in N \mid x \cdot D \subseteq D\}$ is a Cartan subgroup of N. Since N is nilpotent, $E = N$, and then by (5.16)(ii), all left algebraic group structures on N are of the form $D' = x \cdot D = D$, so D is unique. It is immediate that D is also a Hopf subalgebra of $R(N)$.

The description of the subalgebra $R_u[K; E]$ of the algebra $R(K)$ of representative functions on the simply-connected solvable analytic group K on which the Cartan subgroup E acts unipotently given by (5.9) requires in principle knowledge of the algebra $R(K)$ as an E-module. Since we saw subsequently (5.13) that $R_u[K; E]$ is a left algebraic group structure on K and, hence, finitely-generated, it should be possible to describe $R_u[K; E]$ directly. It is; in fact, this was the way Hochschild & Mostow originally approached left algebraic group structures, and we will now outline their construction.

Let K be a simply-connected solvable analytic group and let E be a Cartan subgroup of K. Let $\mathcal{K} = \mathrm{Lie}(K)$, $\mathcal{E} = \mathrm{Lie}(E)$ and $\mathcal{N} = [\mathcal{K}, \mathcal{K}]$. Choose a basis $x_1, \ldots, x_n$ of $\mathcal{K}$ such that $x_1, \ldots, x_m$ is a basis of $\mathcal{N}$, $x_{l+1}, \ldots, x_m$ is a basis of $\mathcal{N} \cap \mathcal{E}$ and $x_{l+1}, \ldots, x_n$ is a basis of $\mathcal{E}$. Let $e : \mathcal{K} \to K$ be given by $e(\sum c_i x_i) = \exp(\sum_1^m c_i x_i) \exp(c_{m+1} x_{m+1}) \cdots \exp(c_n x_n)$. Then e is an analytic bijection; in particular, given $k \in K$ there are unique complex numbers $u_i(k), i = 1, \ldots, n$ such that $e(\sum u_i(k) x_i) = k$. The functions

$u_1, \ldots, u_n$ are then analytic functions on the group K. We will show that for every $x \in K$ and every j, $u_j \cdot x$ is a polynomial in $u_1, \ldots, u_n$.

For $j > m$, this is trivial, since then $u_j \in \mathrm{Hom}(K, \mathbb{C})$ so $u_j \cdot x = u_j(x) + u_j$. Thus, we take $j \leq m$. Using the bijection e, we can reduce the problem to two cases, namely (1) $u_j \cdot \exp(s)$ for $s \in \mathcal{N}$ and (2) $u_j \cdot \exp(c_k x_k)$ for $k > m$. For case (1), we use the fact that $\mathcal{N}$ is nilpotent as a Lie algebra: there are polynomial functions $f_1, \ldots, f_m$ such that $\exp(\sum_1^m c_i x_i) \exp(\sum_1^m d_i x_i) = \exp(\sum_1^m f_i(c_1, \ldots, c_n, d_1, \ldots, d_m) x_i)$. Now if $s = \sum_1^m c_i x_i$ and $t = \sum_1^m d_i x_i$ are elements of $\mathcal{N}$, and $y = \exp(a_{m+1} x_{m+1}) \cdots \exp(a_n x_n)$, then $u_j \cdot \exp(s)(\exp(t)y) = f_j(c_1, \ldots, c_m, d_1, \ldots, d_m)$, so $u_j \cdot \exp(s) = f_j(c_1, \ldots, c_m, u_1, \ldots, u_m)$. Thus, $u_j \cdot \exp(s)$ is a polynomial in $u_1, \ldots, u_m$ and we have established case (1). For case (2), we use the fact that the Lie algebra $\mathcal{E}$ is nilpotent: there are polynomial functions $g_1, \ldots, g_m$ such that $\exp(d_k x_k) \exp(c_{m+1} x_{m+1}) \cdots \exp(c_n x_n) = \exp(\sum_1^m g_i(d_k, c_{m+1}, \ldots, c_n) x_i) \exp(c_{m+1} x_{m+1}) \cdots \exp(cc_k + d_k)x_k) \cdots \exp(c_n x_n)$. To compute $u_j \cdot \exp(d_k x_k)$, we let $p = \exp(d_k x_k)$, $s = \sum_1^m a_i x_i$, $q = \exp(s)$, $r = \exp(c_{m+1} x_{m+1}) \cdots \exp(c_n x_n)$ and r' the product r with the kth factor replaced by $\exp((c_k + d_k)x_k)$. We also let T denote the linear endomorphism $e^{\mathrm{ad}(d_k x_k)}$ of $\mathcal{N}$. We want to write pqr to evaluate $u_j(pqr)$. Now, $pqr = (pqp^{-1})pr$, so from the above we have $pqr = \exp(T(s)) \exp(\sum_1^m g_i(d_k, c_{m+1}, \ldots, c_n) x_i) r'$. To combine the first two factors, we use the functions f_i introduced in the proof of part (1): if $T(x_i) = \sum_1^m d_{ji} x_j$ then the product of those factors is $\exp(\sum_1^m f_i(a_1', \ldots, a_m', b_1', \ldots, b_m') x_i)$ where $a_i' = \sum d_{ij} a_j$ and $b_i' = g_i(d_k, c_{m+1}, \ldots, c_n)$. Thus, $u_j(pqr) = f_j(a_1', \ldots, a_m', b_1', \ldots, b_m')$ so $u_j \cdot \exp(d_k x_k) = f_j(\sum_1^m d_{1i} u_i, \ldots, \sum_1^m d_{mi} u_i, g_1(d_k, u_{k+1}, \ldots, u_n), \ldots, g_m(d_k, u_{m+1}, \ldots, u_n))$ and is, thus a polynomial in $u_1, \ldots, u_n$.

The analytic isomorphism $e : \mathcal{K} \to K$ shows that the algebra D of analytic functions on K generated by $u_1, \ldots, u_n$ is a polynomial ring in $u_1, \ldots, u_n$ and that the maximal ideals of D correspond to points of K. We also know, from the preceding paragraph, that if $x \in K$, $D \cdot x \subseteq D$. To see that D is a left algebraic group structure on K, we need only show futher that $D \subseteq R(K)$; i.e., that each u_i is a representative function.

For $i > m$, this follows from the formula $u_i \cdot x = u_i(x) + u_i$ used above: $[u_i]_r = \langle 1, u_i \rangle$. For $i \leq m$, we choose a faithful representation $p: K \to \mathrm{GL}_n\mathbb{C}$, for suitable n, and let $\bar{K}$ be the Zariski-closure of $p(K)$. The algebraic group $\bar{K}$ is solvable and equals, as a variety, the product of its unipotent radical U and a maximal torus T. Thus, any polynomial function on U is also a polynomial function on K (precede it with projection). In particular, this means that given any linear functional $f: \mathrm{Lie}(U) \to \mathbb{C}$, the function $f \circ \exp^{-1}: U \to \mathbb{C}$ gives a polynomial function on K (we recall that U being unipotent implies that $\exp: \mathrm{Lie}(U) \to U$ is an isomorphism of varieties). Now $(\bar{K}, \bar{K}) = p(K, K)$ is contained in U, and $\mathrm{Lie}(K, K) = \mathcal{N}$, so the functions u_i for $i < m$ are of the form $(f \circ \exp^{-1}) \circ p$ for suitable functionals f on $\mathrm{Lie}(U)$ and, hence, representative.

We have now concluded that D is a left algebraic group structure on K. To conclude that actually $D = R_u[K; E]$ we must further show that $a \cdot D \subseteq D$ for all $a \in E$. Let $a = \exp(\sum_{l+1}^n d_i x_i)$, let $t = \sum_{i=1}^m e_i x_i$, let $p = \exp(t)$ and let $q = \exp(c_{m+1} x_{m+1}) \cdots \exp(c_n x_n)$. We are going to compute pqa. Since $\mathscr{E}$ is nilpotent, there are polynomial functions $h_1, \ldots, h_m$ such that, if $s = \sum h_i(d_{l+1}, \ldots, d_n, c_{m+1}, \ldots, c_n) x_i$, $qa = \exp(s) \exp((c_{m+1} + d_{m+1}) x_{m+1}) \cdots \exp((c_n + d_n) x_n)$. We call the second factor in this last product q', so now $pqa = (p \exp(s))q'$. Now if $j \leq m$, this calculation implies that $u_j(pqa) = f_j(c_1, \ldots, c_m, h_1(d_{l+1} \cdots), \ldots, h_m(\ldots, c_n))$, so we have that $a \cdot u_j = f_j(u_1, \ldots, u_m, h_1(d_{l+1}, \ldots, d_n, u_{m+1}, \ldots, u_n), \ldots, h_m(d_{l+1}, \ldots, u_n))$, so $a \cdot u_j \in D$. Of course, if $j > m$ we have $a \cdot u_j = u_j + u_j(a)$ so $a \cdot u_j \in D$ in this case also. Thus, we have $a \cdot D \subseteq D$ for all $a \in E$. This shows that E is contained in the Cartan subgroup associated to D by (5.16)(i) and, since there are no containments among Cartan subgroups, we have shown that $D = R_u[K; E]$.

The complete and explicit description of left algebraic group structures available in the simply-connected solvable case is not possible in general. However, with the above results as a guide (especially (5.6) and (5.16)) we can still obtain a 'linearization' of basal left algebraic group structures. We proceed as follows: from (5.5), we obtain a nucleus, and from (5.6) a Cartan subgroup of that nucleus, then we show that the nucleus, its Cartan subgroup, and an appropriate reductive group determine the left algebraic group structure. We produce the Cartan subgroup and the reductive group using the procedure of (5.14): they form the 'M' group of that lemma.

Definition 5.18 Let G be an analytic group and let D be a left algebraic group structure on G. The *core* of D is the set of all x in G such that $x \cdot D \subseteq D$.

Lemma 5.19 Let G be an analytic group and let D be a left algebraic group structure on G. Then the core of D is a subgroup of G.

Proof. The only nontrivial point to be seen is that if x is in the core of D, so is x^{-1}. This will be clear if we can see that $x \cdot D \subseteq D$ implies $x \cdot D = D$. Now, by (5.12), $x \cdot D$ is also a left algebraic group structure on G. Let B be the smallest Hopf subalgebra of $R(G)$ containing $x \cdot D$. Since $x^{-1} \cdot B \subseteq B$, we also have $D \subseteq B$, so B is also the smallest Hopf subalgebra containing D. It follows from (5.1) and (4.14) that there are tori T and T' in $G(B)$ such that $D = \{f \in B \,|\, t \cdot f = f$ for all $t \in T\}$, $T = \{y \in G(B) \,|\, y \cdot f = f$ for all $f \in D\}$, $x \cdot D = \{f \in B \,|\, t \cdot f = f$ for all $t \in T'\}$ and $T' = \{y \in G(B) \,|\, y \cdot f = f$ for all $f \in x \cdot D\}$. Since $x \cdot D \subseteq D$, we have $T \subseteq T'$. On the other hand, the descriptions of T and T' show that $x^{-1}T'x \subseteq T$, so $T = T'$ and $x \cdot D = D$.

We can actually say a good deal more about the core as a subgroup:

Proposition 5.20 Let G be an analytic group, let D be a basal left algebraic structure on G and let H be the core of D. Let B be the smallest Hopf subalgebra of $R(G)$ containing D, let $T = \{x \in G(B) \,|\, x \cdot f = f$ for all f in $D\}$, and let $C = \{x \in G(B) \,|\, xt = tx$ for all t in $T\}$. Then $C = e_B(H) \times T$. Moreover, H is a connected analytic subgroup of G which contains a maximal reductive subgroup of G.

Proof. Let P be a maximal reductive subgroup of G such that $(T, e_B(P)) = e$. Let N be the normalizer of T in $G(B)$, and let U be the unipotent radical of $G(B)$. By (5.4) and (4.19), $G(B) = UT \cdot e_B(P)$ (semidirect product) so $N = (N \cap UT) \times e_B(P)$. Since $N \cap UT = C \cap UT$ and $e_B(P) \subseteq C$, $N = C$. Thus, by (5.14), $C = \{x \in G(B) | x \cdot D \subseteq D\}$. Since $C = (C \cap e_B(G)) \times T$, and $C \cap e_B(G) = H$ by definition, we have that $C = e_B(H) \times T$. Since $C \cap UT = (C \cap U) \times T$ is connected and Zariski-closed in $G(B)$, and $C = (C \cap UT) \times e_B(P)$, $H = e_B^{-1}(C \cap e_B(G))$ is a connected analytic subgroup of G, containing the maximal reductive subgroup P of G.

In the notation of (5.20), a maximal reductive subgroup P of the core H is a maximal reductive subgroup of G. Since H has a faithful representation, it has a nucleus E and $H = E \cdot P$ with $E \cap P = \{e\}$. In fact, it has a canonical nucleus as we will now see. First, we formalize some notation:

Notation 5.21 Let G be an analytic group, let D be a basal left algebraic group structure on G and let H be the core of D. Let B be the smallest Hopf subalgebra of $R(G)$ containing D, let $T = \{x \in G(B) | x \cdot f = f \text{ for all } f \text{ in } D\}$, let $C = \{x \in G(B) | xt = tx \text{ for all } t \text{ in } T\}$, and let U be the unipotent radical of $G(B)$. Then we call $(G(B), e_B, T)$ the reduced split hull corresponding to D, C the centralizer corresponding to D, and $e_B^{-1}(UT)$ the nucleus corresponding to D.

Proposition 5.22 Let G be an analytic group, let D be a basal left algebraic group structure on G, and let H be the core of D. Then H is an affine algebraic group with coordinate ring $D|H$, the unipotent radical W of H with this structure is a Cartan subgroup of the nucleus K corresponding to D, and $W = H \cap K$.

Proof. Let $(G(B), e_B, T)$ be the reduced split hull corresponding to D and let C be the corresponding centralizer, as in (5.21). We will regard G as a subgroup of $G(B)$ and e_B as inclusion, and denote $G(B)$ by $\bar{G}$. The inclusion $G \subseteq \bar{G}$ induces the bijection $G \rightarrow \bar{G}/T$, and by (5.20) this bijection induces a bijection $H \rightarrow C/T$. Since

T is central in C, this is an isomorphism of groups. The algebraic variety $\bar{G}/T$ has coordinate ring D, so the closed subvariety C/T has coordinate ring $D|(C/T)$, so that $D|H$ is the coordinate ring of H as an algebraic variety. We also have $D|(C/T) = \{f \in (B|C)\,|\,t \cdot f = f$ for all $t \in T\}$, which shows that $D|(C/T)$ is the Hopf algebra of the algebraic group C/T and, hence, H is an algebraic group with coordinate ring $D|H$.

Let P be a maximal reductive subgroup of G with $(T, P) = e$. Then $\bar{G} = U \cdot (TP)$, so $C = (C \cap U) \cdot (TP)$, and the unipotent radical of C is $C \cap U$. Thus, the unipotent radical of C/T is $(C \cap (UT))/T$. Now $U \cdot T = K \cdot T$ and $C \cap (K \cdot T) = (C \cap K) \cdot T$, so under the isomorphism $H \to C/T$, W becomes the subgroup $(C \cap (UT))/T = ((C \cap K) \cdot T)/T$ and, hence, $W = C \cap K$. Also, $C \cap K = (C \cap UT) \cap K$, and $C \cap UT = \{x \in UT\,|\,xt = tx$ for all t in $T\}$. Now by (5.5), (UT, e_B, T) is a split hull of K, so by (5.6) W is a Cartan subgroup of K. Since $C \cap G = H$ and $W = C \cap K$, we have $W = H \cap K$.

It is important to recognize that the algebraic group structure on H in (5.22) does *not* make H an algebraic subgroup of $\bar{G}$ (to use the notation of the proof of (5.22)) but is relative to H as a subvariety of $\bar{G}/T$. For example, when G is simply-connected solvable with $D = R_u[G; E]$ for a suitable Cartan subgroup E of G, then by (5.15) $H = E$ and we also have from (5.10) that $W = E$, but by (5.6) the Zariski-closure of E in $\bar{G}$ is $E \times T$.

If we want to think positively, we should regard the fact that H is a closed subvariety of $\bar{G}/T$ as the assertion that H is an algebraic subgroup of the left algebraic group G structured by D: that is, H is a closed subvariety of G when G is regarded as an affine algebraic variety with coordinate ring D. In the next two lemmas we make this structure more explicit. Our ultimate goal, in (5.28) below, is that the core of a basal left algebraic group structure determines the structure.

Lemma 5.23 Let G be an analytic group, let D be a basal left algebraic group structure on G, and let H be the core of D. Let K be the nucleus corresponding to D (5.21) and let P be a maximal reductive subgroup of G contained in H (5.20). Then K and P are Zariski-closed subsets of the affine variety G with coordinate ring D, and $K \times P \to G$ by $(k, p) \mapsto kp$ is an isomorphism of affine varieties. Moreover, $D|K$ and $D|P$ are left algebraic group structures on K and P.

Proof. In our Notation (5.21), $K = e_B^{-1}(UT)$, and by (5.5) $e_B(K) \cdot T = UT$. We have also $G(B) = UTe_B(P)$ with $e_B(P) \cap UT = \{e\}$. Under the bijection $G \to G(B)/T$, then, K corresponds to $(UT)/T$ and $\bar{G}/T$ is the product of the two affine subvarieties $(UT)/T$ and $e_B(P)$. Thus, K and P are both closed in G and, of course, $K \times P \to G$ is a bijection. To see that it is an isomorphism, it is enough to know that it is a morphism, since we deal with nonsingular characteristic zero varieties. Now $e_B(P)$, being an algebraic subgroup of $G(B)$, is such that $G(B) \times e_B(P) \to G(B)$ by $(x, p) \mapsto xp$ is a morphism. Since $(T, e_B(P)) = e$, this morphism passes to the quotient $(G(B)/T) \times e_B(P) \to G(B)/T$, or $G \times P \to G$. In particular, this means that $K \times P \to G$ is a morphism. Finally, since K is Zariski-closed in G, $D|K$ is a finitely-generated algebra of representative functions on K whose maximal ideals correspond one–one to points of K. If $x \in K$, $D \cdot x \subseteq D$ (since D is a left algebraic group structure on G) so $(D|K) \cdot x \subseteq (D|K)$, and thus $D|K$ is a left algebraic group structure on K. Similarly, $D|P$ is a left algebraic group structure on P.

We know by (5.16) that every left algebraic group structure on the nucleus K of (5.23) is of the form $R_u[K; E]$ for an appropriate Cartan subgroup E of K. We identify this subgroup E, and record some of the further properties of $D|K$, in the next lemma.

Lemma 5.24 Let G be an analytic group, let D be a basal left algebraic group structure on G, and let H be the core of D. Let K be the nucleus corresponding to D (5.21), let P be a maximal reductive subgroup of G contained in H (5.20), and let $W = H \cap K$. If $p \in P$ and $x \in K$ let $a(p)(x) = pxp^{-1}$. Then:

(1) W is a Cartan subgroup of K, $a(p)(W) = W$ for all $p \in P$, and $D|K = R_u[K; W]$;

(2) If $f \in D|K$ and $p \in P$, then $f \circ a(p) \in D|K$ and $\langle f \circ a(p) | p \in P \rangle$ is finite-dimensional.

Proof. W is a Cartan subgroup of K (and a normal subgroup of H) by (5.22). Since $x \cdot D \subseteq D$ for all $x \in H$, $x \cdot (D|K) \subseteq D|K$ for all $x \in H \cap K = W$. By (5.16), $D|K = R_u[K; E]$ for some Cartan subgroup E of K. By (5.15), $E = \{x \in K \mid x \cdot R_u[K; E] \subseteq R_u[K; E]\}$, so $W \subseteq E$ and, hence, $W = E$ by the maximality of Cartan subgroups. Thus, $D|K = R_u[K; W]$. This proves (1). To see (2), we consider the smallest Hopf subalgebra B of $R(G)$ containing D, and let $X = \langle x^{-1} \cdot f \cdot x \mid x \in P \rangle \subseteq B$. Since f is representative on $G(B)$, X is finite-dimensional. Since $P \subseteq H$, and if $x \in H$ then $x \cdot D \subseteq D$ and $D \cdot x \subseteq D$, we have $X \subseteq D$. Then $X|K = \langle f \circ a(p) \mid p \in P \rangle$ is finite-dimensional and contained in $D|K$, so (2) follows.

Remark 5.25 Using (5.23) and (5.24), we can produce a description of D: from (5.23) we have that $K \times P$ is isomorphic to G as varieties, where G has coordinate ring D and K, P are the closed subvarieties with coordinate rings $D|K$ and $D|P$. We know from (5.24)(1) that $D|K = R_u[K; W]$. Since P is reductive and $D|P$ is a left algebraic group structure on P, it follows from (4.11) and (4.7) that $D|P = R(P)$. Thus, $K \times P$ has coordinate ring $(D|K) \otimes (D|P) = R_u[K; W] \otimes R(P)$, and under the isomorphism $K \times P \to G$ this algebra is sent to D. We shall be more explicit about this isomorphism: write $G = KP$ and, for $x \in G$, let $x_K x_P = x$ with $x_K \in K$ and $x_P \in P$. Define $s: R_u[K; W] \to R(G)$ and $t: R(P) \to R(G)$ by $sf(x) = f(x_K)$ and $tf(x) = f(x_P)$. Since $x \mapsto x_P$ is an analytic homomorphism, which is onto, t is well-defined and injective. To know that s is well-defined, we must have that sf is a representative function on G. For $p \in P$, let $a(p)(x) = pxp^{-1}$. Then, if $kp \in G$, with $k \in K$ and $p \in P$, we have $(sf) \cdot (kp) = s((f \cdot k) \circ a(p))$. So (5.24)(2) shows that sf is representative. Now s and t are injections of complex algebras and $s \otimes t: R_u[K; W] \otimes R(P) \to R(G)$ is injective with image D. It follows that to obtain D, we need to know the core H, the corresponding nucleus K, and a maximal reductive subgroup P of the core.

As it turns out, we can do somewhat better: the subgroup $W = H \cap K$ is already enough to determine the left algebraic group structure. The following lemma and proposition explain how.

Lemma 5.26   Let G be an analytic group, let D be a basal left algebraic group structure on G, let H be the core of G, let R be the radical of G, let K be the nucleus corresponding to D and let $W = H \cap K$. Then $K = (R, G)W$. In particular, W determines K.

Proof. Let P be a maximal reductive subgroup of G, and let S be the radical of P. We use script letters to denote Lie algebras: $\mathrm{Lie}(G) = \mathcal{G}$, etc. Now $[\mathcal{R}, \mathcal{G}] = [\mathcal{R}, \mathcal{K}] + [\mathcal{R}, \mathcal{P}] = [\mathcal{R}, \mathcal{K}] + [\mathcal{K}, \mathcal{P}] + [\mathcal{S}, \mathcal{P}]$ (here we use that $\mathcal{G} = \mathcal{K} + \mathcal{P}$ and $\mathcal{R} = \mathcal{K} + \mathcal{S}$). Since $[\mathcal{S}, \mathcal{P}] = 0$, we find $[\mathcal{R}, \mathcal{G}] \subseteq [\mathcal{K}, \mathcal{G}]$. The reverse inclusion is trivial, and so $[\mathcal{R}, \mathcal{G}] = [\mathcal{K}, \mathcal{G}]$. By (5.22), $\mathcal{W}$ is a Cartan sub-algebra of $\mathcal{K}$, so $\mathcal{K} = [\mathcal{K}, \mathcal{K}] + \mathcal{W} = [\mathcal{R}, \mathcal{G}] + \mathcal{W}$ and (5.26) follows.

Proposition 5.27   Let G be an analytic group, let D be a basal left algebraic group structure on G, let H be the core of G and let K be the associated nucleus. Then $H = \{x \in G \mid x(H \cap K)x^{-1} \subseteq H \cap K\}$.

Proof. If A is a subgroup of the group B, we let $N_B(A) = \{x \in B \mid xAx^{-1} \subseteq A\}$ (the normalizer of A in B). By (5.20), H contains a maximal reductive subgroup P of G and $G = KP$. Since $P \subseteq N_G(H \cap K)$, we have $N_G(H \cap K) = N_K(H \cap K)P$. By (5.22), $H \cap K$ is a Cartan subgroup of the (simply-connected) group K, so $N_K(H \cap K) = H \cap K$. Thus, $N_G(H \cap K) = H$, as claimed.

We now can state the main theorem describing basal left algebraic group structures:

Theorem 5.28   Let G be an analytic group.
  (1) (Uniqueness) If D and D' are basal left algebraic group structures on G with cores H and H' and corresponding nuclei K and K', then $D = D'$ if and only if $H \cap K = H' \cap K'$.
  (2) (Existence) Let K be a nucleus of G and let E be a Cartan subgroup of K. Then there is a basal left algebraic group structure on G with corresponding nucleus K whose core meets K in E.

Proof. (1) For the nontrivial implication, assume $H \cap K = H' \cap K'$.

By (5.27) then, $H = H'$, and by (5.26), $K = K'$. Let P be a maximal reductive subgroup of G contained in H. By Remark (5.25), D' is determined by H, K and P so $D = D'$.

(2) As we observed in the remarks following (4.19), the nucleus K is the nucleus corresponding to some basal left algebraic group structure D' on G. Let H' be the core of D' and let $E' = D' \cap K$. Now E' is a Cartan subgroup of K (5.22), so there is $x \in K$ with $E = xE'x^{-1}$. Let P' be a maximal reductive subgroup of G contained in H' (5.20). Then $H' = E'P'$ (and E' is a nucleus of H' by (5.22)), and $xH'x^{-1} = EP$, where $P = xP'x^{-1}$ is also a maximal reductive subgroup of G. By (5.12), $D = x \cdot D'$ is a left algebraic group structure on G and $\{y \in G \mid y \cdot D \subseteq D\} = \{y \in G \mid x^{-1}yx \cdot D' \subseteq D'\} = xH'x^{-1}$ (5.18), so D has core $H = EP$. (Since the core of D contains a maximal reductive subgroup of G, D is basal.) It is clear that K is the nucleus corresponding to D and that $H \cap K = E$.

There are a number of ways to view (5.28). The essential point is that the subgroup $H \cap K$ of the core determines the left algebraic group structure. This subgroup can be thought of either as a Cartan subgroup of K or as the unipotent radical of the algebraic group structure on H. Either way, once one knows it the left algebraic group structure is pinned down.

Also, (5.28) tells us how to classify basal left algebraic group structures: they are first divided up by corresponding nuclei, and then structures with the same nucleus are further divided by the Cartan subalgebras. (This makes reductive subgroups seem unnecessary: they are actually taken care of automatically by the 'basal' requirement.) One may even wonder why nuclei need be mentioned at all, in the light of (5.26). The problem is that it is difficult, without selecting a maximal reductive subgroup, to say directly which groups occur as Cartan subgroups of nuclei. In fact, the simple connectivity of nuclei makes them useful. We will see next how to completely 'linearize' the study of left algebraic group structures by translating the problem to Lie algebras. Because of (5.28)(2), all we really need to see is how to tell which ideals are Lie algebras of nuclei, and this turns out to be quite simple.

Theorem 5.29 Let G be an analytic group having a faithful representation and let P be a maximal reductive subgroup of G. An analytic subgroup K of G is a nucleus if and only if $\mathrm{Lie}(K)$ is an ideal of $\mathrm{Lie}(G)$ which is a direct sum complement to $\mathrm{Lie}(P)$.

Proof. Let $\mathscr{G} = \mathrm{Lie}(G)$, $\mathscr{P} = \mathrm{Lie}(P)$ and $\mathscr{R} = \mathrm{rad}(\mathscr{G})$. It is clear that the Lie algebra of a nucleus is an ideal of $\mathscr{G}$ and a direct sum complement to $\mathscr{P}$. Conversely, let I be such an ideal. Let L be any nucleus of G and let $J = \mathrm{Lie}(L)$. Now J is solvable and so contained in $\mathscr{R}$, and $\mathscr{R} = J + \mathrm{rad}(\mathscr{P})$. Also, $\mathscr{G}/(\mathrm{rad}(\mathscr{P}) + I) = \mathscr{P}/\mathrm{rad}(\mathscr{P})$ is semisimple, so $J + \mathrm{rad}(\mathscr{P}) = \mathscr{R}$ is contained in $\mathrm{rad}(\mathscr{P}) + I$. Since $\dim(I) = \dim(J)$ and $\mathrm{rad}(\mathscr{P}) \cap I = 0$, we conclude that $\mathscr{R} = \mathrm{rad}(\mathscr{P}) + I$ and, in particular, I is solvable.

Now let K be the analytic subgroup of G with $\mathrm{Lie}(K) = I$. Then K, as a subgroup of G, also has a faithful representation and, hence, a nucleus, say K_0. Let T be a maximal reductive subgroup of K with $K = K_0 T$ and $K_0 \cap T = \{e\}$. To see that K is simply-connected we will prove that $T = \{e\}$. We choose a split hull $(\bar{G}, f, S)$ of G with $(f(P), S) = \{e\}$. Let $Q = Sf(P)$: Q is a maximal reductive subgroup of $\bar{G}$ and, hence, some conjugate of Q contains $f(T)$, say $f(T) \subseteq xQx^{-1}$. Since $\bar{G} = f(G)S$ and S is central in Q, we can assume $x = f(y)$ for some $y \in G$. Let $T_0 = y^{-1}Ty$ (this is also a maximal reductive subgroup of K, as K is normal in G since I was an ideal of $\mathscr{G}$). Then, $f(T_0) \subseteq Q \cap f(G) = f(P)$, so $T_0 \subseteq P$. Since $I \cap \mathscr{P} = \{0\}$, $K \cap P$ is discrete (in K) and $T_0 \subseteq K \cap P$. So $T_0 = \{e\}$ and $K = K_0$ is simply-connected. We have also noted that K is solvable and normal. To conclude that K is a nucleus, we must see further that K is closed in G.

We first make some simpifying reductions. It is enough to see that K is closed in the radical of G, so we can assume G is solvable. Then P is a torus and $(G, G) \subseteq K$. Since (G, G) is closed in G, K will be closed if $K/(G, G)$ is closed in $G/(G, G)$. So we can assume G is abelian. Thus, $\exp: \mathscr{G} \to G$ is a surjective analytic homomorphism whose kernel is contained in $\mathscr{P}$. Since $\mathscr{G} = I + \mathscr{P}$ (as Lie algebras), $G = K \times P$ (as Lie groups) and, hence, K is closed and, thus, a nucleus of G.

With (5.29) available, we can see how to construct left algebraic group structures on the analytic group G: first, select a maximal reductive subgroup of G, then use (5.29) to obtain the Lie algebras of nuclei of G, and then use (5.25) to explicitly produce the left algebraic group structures, using the construction following (5.17). We thus have, as promised at the beginning of this chapter, reduced the problem of obtaining the decompositions $R(G) = A[Q]$ (where A is a basal left algebraic group structure on G and $Q = \exp(\mathrm{Hom}(G, \mathbb{C}))$) to internal problems in $\mathrm{Lie}(G)$.

One final point remains to be explained: in the construction outlined in the preceding paragraph, we start with a nucleus K of G and then choose a Cartan subgroup E of K and a maximal reductive subgroup P of G normalizing E. In our original construction of split hulls in (3.16), we decompose K somewhat differently: we choose nilpotent analytic subgroups N and C of K with $K = NC$ such that a maximal reductive subgroup P of G centralizes C (3.15). The two constructions are related as follows: let $(\bar{G}, f, T)$ be the split hull obtained in (3.16). Then $f(C)$ is also contained in the centralizer of D, of T, so C sits inside $f^{-1}(D \cap f(K))$, which by (5.6) is a Cartan subgroup E' of K. We leave to the interested reader the problem of showing that this Cartan subgroup is uniquely determined by C, using the facts that $\mathrm{Lie}(K) = \mathrm{Lie}(N) + \mathrm{Lie}(C) = [\mathrm{Lie}(K), \mathrm{Lie}(K)] + \mathrm{Lie}(E')$ and that all these algebras are nilpotent, from which one can deduce that $\mathrm{Lie}(E') = \{X \in \mathrm{Lie}(K) \,|\, \mathrm{ad}(Y)$ is nilpotent on X for all Y in $\mathrm{Lie}(C)\}$.

Summary of results of Chapter 5

A split hull $(\bar{G}, f, T)$ of the analytic group G is *reduced* if the center of G meets T trivially. Then there is a one–one correspondence between reduced split hulls and left algebraic group structures on G. The reduced split hull associated to a basal left algebraic group structure determines a nucleus K of G and a reduced split hull (hence, left algebraic group structure) on the simply-connected solvable analytic group K.

The left algebraic group structures on a simply-connected solvable analytic group K are parameterized by the Cartan subgroups of K: to the Cartan subgroup E corresponds the left algebraic group structure $R_u[K;E]$ consisting of all representativ functions on K on which E is unipotent; E is recovered from $D = R_u[K;E]$ as $E = \{x \in G \mid x \cdot D \subseteq D\}$.

In general, if D is a basal left algebraic group structure on the analytic group G, the *core* of D is $\{x \in G \mid x \cdot D \subseteq D\}$. This core H is an analytic subgroup of G, $D|H$ is an algebraic group structure on H whose unipotent radical W is $H \cap K$, where K is the nucleus corresponding to D; W is a Cartan subgroup of K, and determines K uniquely. It also determines D: H is the normalizer of W in G, and there is a maximal reductive subgroup P of G contained in H such that $D = R_u[K;W] \otimes R(P)$. Moreover, every pair consisting of a nucleus of G and a Cartan subgroup of that nucleus arises from some basal left algebraic group structure.

Finally, the nuclei of the analytic group G are precisely those analytic subgroups whose Lie algebras are ideals of $\mathrm{Lie}(G)$ which are direct sum complements to the Lie algebra of some maximal reductive subgroup of G.

Appendix
Commutative analytic groups and characters

This appendix collects, with sketches of proofs, the basic facts about commutative groups, and characters, used in the text. In this appendix, $\mathbb{C}^*$ denotes $GL_1\mathbb{C}$. An analytic group isomorphic to a product of n copies of $\mathbb{C}^*$ is called a torus (of dimension n). We use $\mathrm{diag}(x_1, \ldots, x_n)$ for the diagonal matrix with nonzero entries $x_1, \ldots, x_n$.

(A.1) If A is an $n \times n$ complex matrix with $\exp(A) = I$ then A is diagonalizable and its eigenvalues are integral multiples of $2\pi i$. (This is checked using the Jordan decomposition of A as the sum of a diagonalizable and nilpotent matrix which commute.)

(A.2) Let $\phi: \mathbb{C}^* \to GL_n\mathbb{C}$ be an analytic homomorphism. Then there is an invertible matrix B and integers $d_1, \ldots, d_n$ such that $\phi(x) = B^{-1} \mathrm{diag}(x^{d_1}, \ldots, x^{d_n})B$ for all x in $\mathbb{C}^*$.

Proof. There is a matrix A such that for all x in $\mathbb{C}$, $\phi(e^x) = \exp(xA)$. Then $\exp(2\pi i A) = I$, so by (A.1) there is an invertible B with $BAB^{-1} = \mathrm{diag}(d_1, \ldots, d_n)$ where d_i is integral. Then $\phi(x) = B^{-1} \mathrm{diag}(x^{d_1}, \ldots, x^{d_n})B$.

(A.3) Let $\phi: \mathbb{C}^{*(n)} \to GL_n\mathbb{C}$ be an analytic homomorphism. Then ϕ is algebraic. In particular, the image of ϕ is an algebraic torus. (This is a consequence of (A.2) and the fact that ϕ is the coproduct of analytic homomorphisms from $\mathbb{C}^*$ to $GL_n\mathbb{C}$.)

(A.4) A finite-dimensional analytic module for a torus is semisimple. (This follows from (A.3) and the corresponding assertion to (A.4) for algebraic tori.)

127

(A.5) Let G be a commutative analytic subgroup of $GL_n\mathbb{C}$ and suppose $\text{Ker}(\exp_G)$ spans $\text{Lie}(G)$ as a complex vector space. Then G is an algebraic torus.

Proof. Let $\pi = \text{Ker}(\exp_G)$. We regard $\text{Lie}(G)$ as a subspace of $\text{Lie}(GL_n\mathbb{C})$ and identify the latter with the $n \times n$ complex matrices. If $A \in \pi$, $\exp(A) = I$, so by (A.1), $A = 2\pi i D$ where D is diagonalizable with integral eigenvalues. Thus, we can choose a $\mathbb{C}$-basis of $\text{Lie}(G), D_1, \ldots, D_k$ where D_i is diagonalizable with integral eigenvalues. Define $\phi_i : \mathbb{C}^* \to G$ by $\phi_i(e^x) = \exp(xD_i)$ for $x \in \mathbb{C}$. (ϕ_i is well-defined since $\exp(2\pi i D_i) = I$.) Let $T = \mathbb{C}^{*(n)}$ and let $\phi : T \to G$ be the coproduct of the ϕ_i. Then ϕ induces a surjection on Lie algebras, and since $\exp_G$ is surjective this means ϕ is surjective. Thus, G is an algebraic torus by (A.3).

(A.6) Let G be a commutative analytic subgroup of $GL_n\mathbb{C}$. Then, $G = T \times V$ where V is a vector group and T is a torus.

Proof. Let $\pi = \text{Ker}(\exp_G)$, let L_1 be the $\mathbb{C}$-span of π in $\text{Lie}(G)$ and let L_2 be a vector space complement to L_1 in $\text{Lie}(G)$. Let T be the analytic subgroup of G with Lie algebra L_1 and let V be the analytic subgroup with Lie algebra L_2. Then, T is a torus by (A.5), $\exp_G : L_2 \to V$ is an isomorphism so V is a vector group, and $G = \text{Lie}(G)/\pi = (L_1/\pi) \times L_2 = T \times V$.

(A.7) All additive characters of a torus are trivial.

Proof. An additive character ϕ of the torus T induces a map $T \to GL_2\mathbb{C}$ which sends t to $\begin{bmatrix} 1 & \phi(t) \\ 0 & 1 \end{bmatrix}$ and this must be trivial by (A.3).

(A.8) Let $G = T \times V$ where T is a torus and V is a vector group. Then T is uniquely determined as the kernel of all the additive characters of G.

Proof. Linear functionals on V give additive characters of G when preceded by projection on V; T is the kernel of all these additive characters, hence, of all characters by (A.7).

(A.9) Let G be an analytic group and T a normal torus in G. Then T is central.

Proof. The analytic automorphisms of $T = \mathbb{C}^{*(n)}$ are represented by $n \times n$ invertible matrices of integers: if ϕ is such an automorphism $\phi(x_1, \ldots, x_n) = (x_1^{a_{11}} x_2^{a_{12}} \ldots x_n^{a_{1n}}, \ldots)$. Thus, inner automorphisms by elements of G are represented on T by elements of $\mathrm{GL}_n Z$. But there can be no continuous nontrivial representations of the connected group G in $\mathrm{GL}_n Z$.

(A.10) Let $a \in \mathbb{C}$, $a \neq 0$. Then $\mathbb{C}/Za$ is isomorphic to $\mathbb{C}^*$. (Let $b = 2\pi i/a$. Then Za is the kernel of the analytic surjection $\mathbb{C} \to \mathbb{C}^*$ by sending x to c^{bx}.)

(A.11) Let G be a one-dimensional analytic group and suppose there is an analytic epimorphism $\phi : G \to \mathbb{C}^*$. Then $G = \mathbb{C}$ or $G = \mathbb{C}^*$.

Proof. There is a complex number $c \neq 0$ such that $\phi(\exp_G(t)) = e^{ct}$ for all $t \in \mathrm{Lie}(G) = \mathbb{C}$. Let $\pi = \mathrm{Ker}(\exp_G)$ and let $b = 2\pi i/c$. Then π is contained in Zb, so $\pi = Za$ where a is an integral multiple of b. If $a = 0$, $G = \mathbb{C}$ and, if $a \neq 0$, $G = \mathbb{C}/Za$ is isomorphic to $\mathbb{C}^*$ by (A.10).

(A.12) Let G be an analytic group, and let T be a normal torus in G such that $\bar{G} = G/T$ is also a torus. Then G is a torus.

Proof. By induction on the dimension of $\bar{G}$ we may assume $\bar{G} = \mathbb{C}^*$. Let G act on $\mathrm{Lie}(G)$ via the adjoint representation. Since T is central in G, (A.9), the adjoint action factors through $\bar{G}$, so by (A.4) $\mathrm{Lie}(G)$ is semisimple for the adjoint action. The submodule $\mathrm{Lie}(T)$ is then complemented, and a complement is an ideal I isomorphic to $\mathrm{Lie}(\bar{G})$. Thus, $\mathrm{Lie}(G) = \mathrm{Lie}(T) \oplus I$ is the sum of two abelian ideals, so $\mathrm{Lie}(G)$ and G are abelian. The exponential maps carry the exact sequence $0 \to \mathrm{Lie}(T) \to \mathrm{Lie}(G) \to \mathrm{Lie}(\bar{G}) \to 0$ onto the exact sequence $1 \to T \to G \to \bar{G} \to 1$ so we get an exact sequence of kernels $1 \to \mathrm{Ker}(\exp_T) \to \mathrm{Ker}(\exp_G) \to \mathrm{Ker}(\exp_{\bar{G}}) \to 1$. Now $\mathrm{Ker}(\exp_{\bar{G}})$ is infinite cyclic, and we can choose $a \in \mathrm{Ker}(\exp_G)$ mapping onto a generator of $\mathrm{Ker}(\exp_{\bar{G}})$. Then a spans an ideal I of

Lie(G). Let H be the analytic subgroup of G with Lie algebra I. Then $H \to \bar{G}$ is onto, and $H \neq \mathbb{C}$ since $a \in \text{Ker}(\exp_H)$. By (A.11), $H = \mathbb{C}^*$, and it follows that $\mathbb{C}^* = H \to \bar{G} = \mathbb{C}^*$ must be given by an nth power. In particular, the kernel of this map, namely $H \cap T$, is finite. Also $T \times H \to G$ by multiplication is surjective, with kernel isomorphic to the finite group $T \cap H$; $T \times H$ is an algebraic torus, and so is every quotient of its by a finite subgroup, so G is a torus.

(A.13)   Let G be an algebraic group. Every additive analytic character of G is algebraic.

Proof. We may assume G is abelian with maximal torus T. Every additive character vanishes on T (A.7) so factors through the vector group G/T, and an additive character of G/T is a linear functional, hence, algebraic.

Notes

This list gives the sources of the main concepts and results presented in the book. The numbers in square brackets refer to the bibliography.

Chapter 1

(1.7): [4; p. 385]

(1.8): [4; p. 385]: Grothendieck denotes $\text{Aut}_\otimes(\text{Mod}(G))$ by $Cl_C(G)$. Our notation is chosen to reflect the dependence of the group of tensor automorphisms on the category $\text{Mod}(G)$ rather than on the group G.

Example E: [16; p. 1150]: Hochschild & Mostow show that these groups have isomorphic algebras of representative functions. The explicit equivalence of categories presented here is new.

Chapter 2

(2.2): [7; p. 497]

(2.4): [7; Prop. 2.1, p. 497]

(2.21): See [28] for the definitions of Hopf algebra, comultiplication, counit and antipode.

(2.34): [7; p. 501]

(2.37): This theroem is an elaboration of an idea of Alex Lubotzky's.

Chapter 3

(3.5): [6; Theorem 4.6, p. 225]

(3.6): [10, p. 113]

(3.7): [9; Theorem 3.6, p. 95]. The proof here is based on [22; Theorem 2, p. 393].

(3.12): [23; p. 304]

(3.16): This theorem is a finite-dimensional specialization of [10; Theorem 6.1, p. 127]. The proof here is based on [22; Theorem 1, p. 390].

(3.17): [23; Theorem 1, p. 305]

(3.18): [23; Theorem 5, p. 307]

(3.23): [6; Theorem 4.3, p. 223]

(3.24): [20; Theorem 5, p. 876]

(3.26): [7; Theorem 7.1, p. 522]

(3.27): [6; Theorem 4.6, p. 225]

Chapter 4

(4.7): [10; Theorem 3.1, p. 118]

(4.8): [10; Theorem 6.1, p. 127]

(4.10): [9; Theorem 5.3, p. 99]

(4.11): [9; Theorem 5.4, p. 99]

(4.13): [18; p. 1052]

(4.18): [10; p. 116]: 'Basal' is a contraction of 'normal basic'; what we call basal left algebraic group structures Hochschild & Mostow call normal basic subalgebras.

Chapter 5

(5.2): [23; p. 307]

(5.6): [19; Cor. 1.9, p. 172]

(5.16): [19; Theorem 1.10, p. 172 and 19; Theorem 1.11, p. 172]

(5.18): [18; p. 1047]

(5.19): [18; Cor. 1.5, p. 1047]

(5.28): [19; Theorem 2.3, p. 174]

(5.29): [24; Lemma 1.1, p. 282]

Bibliography

[1] Barut, A. O. & R. Raczka. *Theory of Group Representations and Applications*, PWN–Polish Scientific Publishers, Warsaw, 1977.

[2] Bourbaki, N. *Lie Groups and Lie Algebras, Part I*, Addison-Wesley, Reading, Mass., 1975.

[3] Chevalley, C. *Theory of Lie Groups, I*, Princeton University Press, Princeton, N.J., 1946.

[4] Grothendieck, A. 'Representations lineaires et compactification profinie des groupes discrets,' *Man. Math.* **2** (1970), 375–96.

[5] Harish-Chandra. 'Lie algebras and the Tannaka duality theorem,' *Ann. Math.* **51** (1950), 299–330.

[6] Hochschild, G. *The Structure of Lie Groups*, Holden-Day, San Francisco, 1965.

[7] Hochschild, G. & G. D. Mostow. 'Representations and representative functions of Lie groups,' *Ann. Math.* **66** (1957), 495–542.

[8] Hochschild, G. & G. D. Mostow. 'Representations and representative functions of Lie groups, II,' *Ann. Math.* **68** (1958), 295–313.

[9] Hochschild, G. & G. D. Mostow. 'Representations and representative functions of Lie groups, III,' *Ann. Math.* **70** (1959), 85–100.

[10] Hochschild, G. & G. D. Mostow. 'On the algebra of representative functions of an analytic group,' *Amer. J. Math.* **83** (1961), 111–36.

[11] Hochschild, G. & G. D. Mostow. 'On the algebra of representative functions of an analytic group, II,' *Amer. J. Math.* **86** (1964), 869–87.

[12] Hochschild, G. & G. D. Mostow. 'Affine embeddings of complex analytic homogeneous spaces,' *Amer. J. Math.* **87** (1965), 807–39.

[13] Hochschild, G. & G. D. Mostow. 'Representative algebraic structures on complex analytic homogeneous spaces,' *Amer. J. Math.* **88** (1966), 847–66.

[14] Hochschild, G. & G. D. Mostow. 'Deformation of affine embeddings of complex analytic homogeneous spaces,' *Amer. J. Math.* **88** (1966), 244–57.

[15] Hochschild, G. & G. D. Mostow. 'Pro-affine algebraic groups,' *Amer. J. Math.* **91** (1969), 1127–40.

[16] Hochschild, G. & G. D. Mostow. 'Complex analytic groups and Hopf algebras,' *Amer. J. Math.* **91** (1969), 1141–51.

[17] Magid, A. 'Left algebraic groups,' *J. of Algebra* **35** (1975), 253–72.

[18] Magid, A. 'Analytic left algebraic groups,' *Amer. J. Math.* **99** (1977), 1045–59.

[19] Magid, A. 'Analytic left algebraic groups, II,' *Trans. Amer. Math. Soc.* **238** (1978), 165–77.

[20] Magid, A. 'Analytic subgroups of affine algebraic groups,' *Duke J. Math.* **44** (1977), 875–82.

[21] Magid, A. 'Analytic subgroups of affine algebraic groups, II,' *Pacific J. Math.* **86** (1980), 145–54.

[22] Magid, A. 'Faithfully representable analytic groups,' *Algebraic Geometry Proceedings Copenhagen 1978*, Lecture Notes in Mathematics No. 732, Springer-Verlag, New York, 1979.

[23] Magid, A. 'Representative functions on simply-connected solvable groups,' *Amer. J. Math.* **102** (1980), 303–19.

[24] Magid, A. 'Moduli for analytic left algebraic groups,' *Trans. Amer. Math. Soc.* **260** (1980), 281–91.

[25] Nakayama, T. 'Remark on the duality for noncommutative compact groups,' *Proc. Amer. Math. Soc.* **2** (1951), 849–54.

[26] Pontryagin, L. 'Theory of topological commutative groups,' *Amer. Math.* **35** (1934), 361–88.

[27] Saavedra-Rivano, N. *Catégories Tannakeinnes*, Lecture Notes in Mathematics No. 265, Springer-Verlag, New York, 1972.

[28] Sweedler, M. *Hopf Algebras*, W. A. Benjamin, Inc., New York, 1969.

[29] Tannaka, T. 'Uber den dualitätssatz der nichtkommutativen topologischen Gruppen,' *Tôhoku. Math. J.* **45** (1938), 1–12.

Index

Analytic group, 1

Cartan subgroup, 109
Character, ix
  additive, ix
  exponential additive, 84
Comodule, 42
  morphism, 44

Left algebraic group structure, 95
  basal, 98
  core of, 117
Left stable, 36

Mod(G), 16
Module, 15

Nucleus, 60

Proper automorphism, 48

$R(G)$, 36
Reductive, 59
Representation, 15
Representative function, 34
  semisimple, 97
Right stable, 36

Split hull, 65
  reduced, 105

Tensor automorphism, 17

Unipotent action on representative
  function, 110